Australia's
HIDDEN
TREASURES

Australia's
HIDDEN TREASURES

A guide to treasure hunting –
what to look for and how to find it

JEFF TOGHILL

Technical adviser Stan Irwin

ANGUS
& ROBERTSON
PUBLISHERS

ANGUS & ROBERTSON PUBLISHERS

Unit 4, Eden Park, 31 Waterloo Road,
North Ryde, NSW, Australia 2113;
94 Newton Road, Auckland 1,
New Zealand; and
16 Golden Square, London W1R 4BN,
United Kingdom

First published in Australia
by Angus & Robertson Publishers in 1988

National Library of Australia
Cataloguing-in-publication data.

Toghill, Jeff, 1932–
   Australia's hidden treasures.

  Bibliography.
  Includes index.
  ISBN 0 207 15910 6.

  1. Treasure-trove—Australia. 2. Prospecting—
Australia. 1. Irwin, Stan. II. Title.

622'.19'0994

Typeset in 11 pt Meriden by Best-set Typesetter Ltd
Printed in Hong Kong

ACKNOWLEDGEMENTS

Many people helped in the compilation of this book — far too many to mention individually. My thanks go to all, but in particular to:

Tom Budden, "our man" in the north, whose extensive travels and excellent photographs have done much to enhance this book.

Paul and Shirley Hanke, Bob Greig, Marie and Barry Ayers, Robyn Hill, Dede and John Deegan, Dr Bernie Brinsmead, Phil Coleman and Vicki Armstrong, for their fine photographs.

Mike McCarthy and Pat Baker of the Western Australian Maritime Museum for their wonderful co-operation and photographs.

David Barnes, NSW Department of Mineral Resources photographer, for his kind assistance and photographs.

Jacquie and David Ross, of Yarrabin, for their advice and assistance with field work in their locality.

Dell Oulton of Dubbo, for assistance with the "Midnight" story.

Margaret Gietz, of Sydney, for her expert advice on gemstones.

Mosman Country Antiques.

The Geological and Mining Museum, Sydney.

The Mitchell Library and the State Library of New South Wales, Sydney.

Northern Territory Department of Mines and Energy.

Department of Environment and Planning, South Australia.

Department of Industry, Technology and Resources, Melbourne.

Department of Mines, Queensland.

Geological Survey of Western Australia, Perth.

Department of Mines and Energy, South Australia.

Department of Mineral Resources, New South Wales.

The Sapphire and Opal Centre, Sydney.

G & J Gems, Sydney.

Most of all, my thanks to my technical adviser, Stan Irwin, for his tireless assistance with the great volume of research work involved in this book.

CONTENTS

INTRODUCTION

Treasure hunting is alive and well in every city, town and village across the nation. It is alive in the hills and the valleys, the rivers and the creeks, the coastal waters and the arid deserts. Far from being an anachronism in this world of computerised games and electronic entertainment, treasure hunting, in one form or another, is actively pursued by a large part of Australia's population.

But, contrary to popular belief, treasure is not just cash, jewellery or bullion. Films and TV documentaries portray treasure hunting as an operation usually involving a historic wreck or site, which results in the recovery of piles of old coins, gold ornaments or chests of jewellery. Nothing could be further from the truth, for while an occasional great treasure is discovered, such as the 1987 Tennant Creek gold reef, the odds against such a find are far greater than those of winning a major lottery. Most treasure hunters, while always stimulated by the possibility of a lucky strike, find sufficient incentive in the non-tangible rewards provided by the excitement of the hunt, the therapeutic effect of a relaxing pastime or the simple pleasure of a family activity.

Treasure, like beauty, is in the eye of the beholder, and literally anything can be a treasure-trove. The throw-outs of a household clean-up are at one end of the treasure-hunting spectrum, seductively beautiful opals at the other. In between lies a wealth of treasure, providing activities to suit a wide variety of treasure hunters — activities which can, to some, become addictive and an all-consuming passion, but to others just provide a pleasant holiday interest involving the whole family.

Hunting for treasure is both physically and mentally relaxing. There is nothing more certain to relieve the pressure on jangled city nerves than sitting in dappled sunlight on the edge of a country creek, fingering specks of gold or sapphires from the pay-dirt of a panning dish. Even without the reward of an occasional flash of gold or sparkle of blue, the exercise is probably more effective than a hundred visits to a psycho-analyst. Indeed, if the tranquil country atmosphere, the chatter of birdsong and the gentle movement of trees and creek could be captured and bottled, it would eliminate for all time the need for tranquillising drugs.

At the same time, treasure hunting is stimulating and educational. The chance of obtaining a bargain, or the chance of making a rich strike, creates a stimulating incentive, while researching and tracking down the treasure is a good educational exercise. Especially when it involves delving into fascinating subjects across a wide spectrum of interests, such as the value of old paintings, photographs and artefacts, or the geological history of the earth's crust which led to deposits of gold and precious stones.

Since a great deal of treasure involves items from the past, the historical side of treasure hunting can also be both educational and interesting. A favourite location for gold fossicking is around the old mines that played a prominent part in establishing Australia on the world map. Research before commencing field work in such areas can bring to light fascinating aspects of life in the heady days of the gold rushes. Such information not only makes for interesting reading but also gives an added facet to visiting and working an old goldfield.

Travel is another plus, for there is no better way to see the countryside than when looking for treasure. Whether the search involves ghost towns, old properties, derelict mines or mountain streams, most are far from the major cities and are often off the beaten track. Australians can be very surprised at the magnificence of their own country when they leave the highway and plunge into the non-tourist regions of the outback.

These are just some of the lesser-known facets of treasure hunting. There are many more. This book is designed to just push the gates ajar, for beyond is a world of interest, excitement, pleasure and reward that could not be described in a dozen volumes the size of this.

A real treasure in anyone's language: a beautifully formed crystallised gold nugget, found near Nundle, New South Wales.
PHOTO DAVID BARNES, COURTESY NSW DEPARTMENT OF MINERAL RESOURCES.

CHAPTER ONE

TREASURE HUNTING
HOW, WHY & WHERE

TREASURE, TREASURE, EVERYWHERE

Treasure has many guises. In its more traditional form it is a pile of old coins or valuables dug up from beneath the soil or retrieved from the bottom of the sea. It can be gold in its native state or refined to create a valuable adornment. It may be gemstones, uncut and still dulled by the encasement of host rock, or cut and polished to sparkling brilliance. But not all treasure is traditional and not all lies beneath the ground or the sea. Some treasure is found in the most unlikely locations, often covered by nothing more than a layer of dust. Some is not concealed even by a layer of dust but stands open and obvious to all, yet unseen except by discerning eyes.

Locating and recovering such treasure is not done with electronic metal detectors or scuba-diving gear, nor are picks, shovels, sieves or panning dishes required. The results may not be as spectacular as discovering a gold nugget or unearthing a valuable gemstone, but the search can prove every bit as exciting, and success can be both aesthetically and financially rewarding.

In a world of wildly fluctuating finances, the value of even mundane,

Old books are among the more common treasures found in attics and storerooms. This picture shows pages from the *Picturesque Atlas of Australasia*, published in 1886.

Opposite: Undoubtedly the most spectacular of all treasures, opals are within the reach of even the amateur fossicker. One of the most magnificent black opals taken from the Lightning Ridge field.
PHOTO DAVID BARNES, COURTESY NSW DEPARTMENT OF MINERAL RESOURCES. OPALS COURTESY SAPPHIRE AND OPAL CENTRE.

everyday items changes rapidly. Money itself is perhaps the best example. Almost as it leaves the mint, a dollar coin begins to lose value due to inflation and other monetary pressures. Yet fifty years hence, that dollar coin may be worth several hundreds, even thousands, of dollars as a collector's item. A 1930 Australian penny, in good condition, is worth about $50,000 today on the collectors' market, only 57 years since it was minted! Much the same applies to many items of everyday life — postage stamps, books, furniture, paintings and clocks, to mention but a few. Even old photographs can be valuable if the subject matter they contain, or the photograph itself, has historical significance.

Old is beautiful in many collectors' eyes, and since Australia has been settled for only two hundred years, old may be relatively recent. The first edition of Matthew Flinders' journal, only 174 years old, could be as valuable to an Australian collector as a twelfth-century tapestry to a European.

So where are these treasures to be found? Certainly, in places such as antique shops. But by then they have been recognised as treasures and priced accordingly. The secret of real treasure hunting is to search out and find such trophies while they are still "lost", long before they have been discovered by experts and recognised for their true value. A treasure is only a treasure when it is obtained at a fraction of its real worth.

The traditional "Grandma's attic" has always been a favourite hunting ground for treasure seekers. Since attics are something of an anachronism in Australia today, an equivalent source may be an old garden shed, a storeroom under the house or in the roof, or old chests and trunks retained from the days of sea travel. Providing it contains yesterday's cast-offs, almost any storage area is a potential source of treasure. Yet few would-be treasure hunters think of searching their own house first. One experienced treasure seeker, who has unearthed some priceless treasures in the course of his "attic" fossicking, trains his protégés by first having them search their own lounge-room furniture. Few are not surprised at the results of such an exercise which, although not financially rewarding, brings home the message that every home owner could be sitting on a fortune! Extending the search beyond the lounge room to the attic, or whatever area is used for storage, can reveal even greater, and sometimes rewarding, surprises.

Few modern homes have real treasure stowed away, and in any case, sifting through your own discarded rubbish does not have the appeal of sifting through someone else's, particularly if that someone else lived in a previous generation.

THE COUNTRY CONNECTION

Although urban homes often contain hidden "attic" treasures, nowhere is there more potential for hitting the jackpot than in the country. For some reason, perhaps because of the lack of council rubbish collections, country folk tend to hoard old stuff more than their city counterparts. When an item is discarded, it is usually either thrown into a barn or a shed, or perhaps the attic, and forgotten. As a result, country houses are often an Aladdin's Cave for treasure hunters. An opportunity to search through an old country homestead can be the start of an exciting and rewarding expedition.

Apart from the discarded trappings of past generations, many of the items still in use have considerable value in the furniture markets of the city. A slab timber kitchen table is one example. Other items worth looking out for are "stone" whisky jars and old bottles, usually found cluttering up old dairies or pantries, or even piled up in the backyard. (See Chapter 10, where some of these items are discussed in greater detail.)

Old biscuit or tobacco tins, with intricate designs, and mugs of all shapes, sizes and vintages are often standing on shelves in sheds or barns where they have been forgotten for decades. In really old homesteads, kitchen utensils, long since superseded by modern equipment, are often cast aside and forgotten until unearthed by an assiduous treasure seeker. Many a fine bowl or pot has been recovered from the back of deep kitchen cupboards in country houses.

The kitchen is only one room in the house and, for the diligent seeker, there can be even greater treasures in the other rooms. Paintings, photographs and plaques often hang on the walls for decades, more

Above: Treasures are to be found everywhere, from beneath the sea to Back o' Bourke. The beauty (and value) of the treasure lies in the eye of the beholder.

Right: A treasure hunter's paradise — a deserted homestead on the Monaro Plains.

because they have become part of the established room decoration than because of any intrinsic or personal value. Yet those same items can fetch a good price in the right bric-a-brac shop in the city. Since they are often of no particular value to the owner, the transaction is good business for everyone. And there is no better business deal than one which leaves everyone happy.

Attic junk may include clothes of a bygone era. Even these have potential for the wily treasure hunter, who will hawk them around city theatrical shops where they are valued for anything from stage costumes to fancy-dress party gear. Cloaks, boots, walking sticks, even wigs, can be turned into cash where they would otherwise have lain in closets or trunks for a further decade or until they disintegrated. As often as not, the owner will be only too pleased to see the end of such accumulated junk, particularly if the bargaining leaves his pockets bulging.

The search should not be limited to the house, for even more discarded stuff will be found in the outhouses. Outdated or worn harness equipment, brasses, carriage lanterns and other relics from the days of horse transport are often left lying around the stable, hanging on hooks or tossed into corners. Horse brasses are a popular decoration in many city

homes, and in recent years all kinds of brassware have seen a re-surgence in popularity. With modern imitation horse brasses flooding the market, the real thing has escalated considerably in value.

Old sheds and barns are a treasure hunter's paradise, for under the straw, hay, chaff and other detritus of stock feeding, lie discarded tins and sacks, often dating back to the turn of the century. Buried beneath decades of dust and grime, bottles, jars and equipment lie on cluttered shelves. Scythes, billhooks and other tools of the pre-machinery era will either be hanging up on pegs on the walls or thrown carelessly into heaps in the yard. These are not priceless treasures in themselves but can be the basis of interesting collections, or items of value in the collectors' market place.

Old homesteads, by virtue of their size, age and probable wealth of past owners, provide the best potential for unearthing worthwhile objects, but treasures, particularly those of historic value, are likely to be found in every type of dwelling from miners' huts to country pubs.

Some experienced treasure seekers like to fossick in derelict buildings. It is necessary to get permission from the owner of the land on which the abandoned building stands, but this is rarely a problem. Metal detectors are useful for locating relics beneath fallen rubble or mounds of soil, avoiding much digging and scraping to locate buried objects. A pair of secateurs and shears is also useful since, in many cases, plant growth covers the ruins and must be cut away before serious exploration can begin. For the most part, such buildings are small, and since they were inhabited only by farm hands and shepherds, the likelihood of valuable finds is not very great. However, items of aesthetic value and sometimes useful collectors' items may be found on these sites.

In the country, backyards are often massive junk heaps, and are always worth close investigation. A recognised authority once commented that the junky backyards of the bush are a paradise for treasure seekers. This is particularly true for those with a knowledge of old motor cars, for there is a derelict car in every second backyard and many are so old they could be classified as vintage. Although the cars themselves are usually of no tangible value, anyone with a knowledge of the market for spare parts of vintage or old model vehicles would not need to travel far to find worthwhile treasure of this type.

There are other sources of treasure in the countryside, although these do not provide quite the excitement and stimulus of fossicking through farmhouses, barns and stables, since half the work has already been done. Nevertheless, exciting bargains can be found in the second-hand stores of most country towns, and in the ''farmhouse'' auctions for which some localities are famous. While the details may differ from country town to country town and from state to state, basically the situation is the same in most country areas in Australia.

Second-hand goods may be found in sheds or in shops that are usually run by some local identity. When times are hard — and they often are in the country, as a result of drought, high interest rates, crop failures or some other disaster — many of the country folk live a hand-to-mouth existence. If work is not available they must sometimes sell personal belongings in order to survive. Then the second-hand shops do a roaring trade, providing both a buy and sell service for locals and creating a rich fossicking ground for treasure seekers and collectors. Of course, shrewd second-hand dealers in some of the larger country towns will know the

This blunderbuss was found lying among discarded bric-a-brac in the outhouse of a Victorian farm.
PHOTO JOHN CLARKE.

value of the items they buy and mark them up accordingly. Few treasures will be found at bargain prices in these shops, but those in the know will move out to the smaller towns and villages where either the dealer is less knowledgeable about the value of his goods, or he needs a quick cash turnover.

Some treasure hunters follow the "farmhouse" auctions. These are auctions of goods held either at a house or homestead or in the local hall of the nearest village. The items auctioned may be from a single property or gathered from the neighbouring countryside. They may be the result of a bankruptcy or a deceased estate, or may be a regular event to which residents of the surrounding regions bring their throw-outs, with the twofold purpose of making a few cents while enjoying a day's fun.

One particularly exciting form of disposing of small items is to lump them together in boxes — a veritable potpourri of second-hand items in each box — and put each box up for auction. Then, for a dollar or two, or whatever the bidding reaches, the buyer may get a range of both useful and useless items; the useful to be retained, the useless to be returned in another box of mixed goodies and put up for auction again. The fun of such an auction lies in the fact that bargains fly thick and fast under the auctioneer's hammer. Quite apart from the possibility of finding treasure, attending one of these auctions is more than rewarding for its entertainment value.

Where country auctions are held regularly, astute treasure hunters get to know the auctioneers' circuits. Where they are held spasmodically, the auction dates are advertised in local papers. Since these are not easily attainable in the big cities, it may be necessary to visit a few country areas where the auctions are held and cultivate acquaintances among the locals to provide a tip-off when an auction is to be held.

The more "up-market" auctions, in which genuine antiques are to be offered, are usually well advertised in trade and city papers in order to attract the big buyers. While it is still possible for experienced amateurs to obtain bargains at such auctions, the competition is stiff, taking away much of the treasure-hunting fun and creating a more serious atmosphere. The bric-a-brac farmhouse auctions, however, are a delight for all comers, not just because of the chance of striking a bargain, but because of the relaxed atmosphere and the wonderful characters that make the auction more of a game than a serious sale.

PREPARING FOR A TREASURE HUNT

Setting out on a treasure hunt is not simply a matter of gathering up the tools, the family and the food, jumping into the car and heading out of town. Preparation always has a lot to do with the success of an expedition, and it must begin early. There are four basic steps required to mount a successful treasure hunt: discovering the clues, researching the project, locating the site, and recovering the treasure.

1. Discovering the clues

The first step varies a great deal according to the type of treasure involved. When fossicking for gold, for example, the only requirement at this stage is to determine where to begin. If you are going to fossick through the workings of an old goldfield, then its location will be known and the only requirement will be to research all available records to find the part of the field most likely to produce good results. Similarly with gemstones; rather than taking pot luck with any old creek or stream, it will be necessary to find an area known to have gemstone deposits and then determine which stream to work. Such information can be obtained from government mining authorities or mining museums, or from fossicking and lapidary clubs.

Other types of treasures may be located as the result of following up clues obtained either by accident or by design. A historic example of this resulted in the first discovery of payable gold in Australia. A professional treasure hunter, Edward Hargraves, had left Australia to search for gold in the United States of America. While working in the goldfields of California, he was struck by the similarity between the rock formations in those goldfields and those in the countryside around Bathurst, New South Wales.

This was a clue, and when he returned to Australia, Hargraves followed it up, searching the creeks near Bathurst among rocks similar to those he had seen in California. It is now a matter of history that on 12 February 1851, Edward Hargraves, together with two companions, found gold at the junction of Summer Hill and Lewis Ponds Creeks, near Bathurst.

Clues can be found in a variety of places, not least of which are history books and journals. For example, it is a well-known fact that while many of the more notorious bushrangers made good profits from their illicit trade, few had the time or opportunity to spend them. Dozens of bushrangers were either killed or captured before they could enjoy the benefits of their activities, and their loot probably still lies hidden somewhere in the bush. How, then, to find that hidden treasure?

Reference libraries are the obvious places to begin, and one of the finest is the Mitchell Library in Sydney, where extensive records of Australian history are kept. However, many libraries keep well-researched and documented books on bushrangers, including such classics as *Wild Colonial Boys* by Frank Clune. From works such as this, a great deal of information, and particularly dates, can be gathered. A good follow-up is to read the newspapers and journals which cover those dates. Gradually a picture is built up of the bushranger and his activities, the areas in which he moved, and perhaps the places he frequented. Much of this research may lead to dead ends, but occasionally one will provide a lead; a clue on which to build towards an expedition.

Clues often turn up by accident. A well-known maritime expert working on a shipwreck in Tasmania in 1984 became puzzled by conflicting details in reports of the ship's movements. Careful research revealed that there were two vessels with the same name and the reports had become confused with time. This was the clue that he followed up diligently, and when eventually the second wreck was located, it proved to have considerably greater historic value than the first.

Clues to treasures can be found much closer to home. Old letters that may have been stowed away in attic trunks can reveal exciting clues about unknown family treasures. Perhaps there was a safe deposit box that Grandad forgot to mention in his will. Or a packet of diamonds Great Aunt hid because she distrusted banks. Or there might be a reference to the original sketch Hans Heysen gave to Great-Great-Grandfather when they worked together in the Barossa Valley.

Prospector Edward Hargraves found his clues in the similarity between these rocks at Lewis Ponds Creek and those of gold-bearing country in California.

2. Researching the project

This is sometimes the most interesting part of treasure hunting, sometimes the most demanding, but always the most important. Much depends on the subject, the availability of research material and the amount of work involved. The incentive to find the missing treasure is usually sufficient stimulus to overcome the more monotonous aspects of the work, but since the chance of locating the treasure can be enhanced by every extra hour of research, the effort is always worth while.

As with finding clues, much information relating to treasures of a general nature can be found in libraries and museums. In our hypothetical case of the bushranger's cache, the library research would concentrate on all known aspects of the bushranger, his illegal activities and his movements. With this information documented, it would be time to get out into the field, particularly to visit museums and historical societies in the neighbourhood where the bushranger was active. Local newspapers are often a good source of information, particularly old clippings which highlight any local activity related to the subject.

The chances of finding a living relative of a bushranger are remote now that more than two generations have passed since the bushrangers' era. However, this angle is worth investigating, for even distant relatives, or relatives of friends, may have relics which will give some clue or confirm information previously obtained.

Fossicking for gold and gemstones can be enhanced by careful research. Information about the operation of a mine, for example, may reveal that it was at its most prolific midway through its operating life. This information can be put to good use when fossicking through the tailings, for the most likely concentration of gold will be found at about mid-depth in the tailings heap. Deeper research may unearth even more specific information, such as the type of gold produced by a specific shaft, and the mound of tailings in use at the time. It takes only a modest amount of intelligence to convert past records into useful present-day information.

Old mining towns are a good place to research information on abandoned gold fields. This is Captains Flat, near the still-productive Araluen and Shoalhaven rivers, New South Wales.

The history of a goldfield can also provide useful information. White miners, in their frantic search for quick riches, were not over-meticulous in their operating methods and would miss fragments of gold in their eagerness to find big nuggets. In areas worked by European miners, therefore, the chances of success when panning gold from the creeks or running a metal detector over mullock heaps are good. The industrious Chinese, by contrast, sifted through every spoonful of soil assiduously, leaving not a trace of colour behind, so the chance of finding gold where the Chinese once worked is less likely.

Of course, this applies only to tailings and mullock heaps, for the creeks, even where the Chinese panned for gold, have since been washed by floods that may have carried gold-bearing silt to replenish the deposits depleted by earlier activities.

Research is particularly important when fossicking for gemstones, since a valuable gem can be by-passed or lost through lack of knowledge. Sapphires, for example, are not always a brilliant blue, as is the popular belief. They can be many shades of blue, yellow, green or violet, and many an inexperienced gem hunter has cast away a valuable stone unaware that it was a sapphire. Since the colour of a stone depends on the locality in which the gems are found, research before a fossicking expedition will indicate the colour of the gemstones likely to be found in a particular district. Information on gemstones can be obtained from state mining authorities, mineral or gem museums and fossicking or lapidary clubs.

It is obvious, then, that research is an important step in mounting a treasure hunt. Apart from influencing the success of the venture, effective research can minimise the amount of work involved both in locating and recovering the treasure. A few hours in a library or museum can mean a saving of many hours in the field.

3. Locating the site

When the research work is completed, the field work begins. Just how much field work is required to locate the treasure will depend on how much was revealed by the research and the nature of the treasure itself. The position of a cache buried in a ghost town, for example, can often be narrowed down to one specific house or backyard if the research has uncovered sufficient material to pinpoint it. By contrast, the exact location of a gold nugget (even if one exists) can never be pinpointed, although reports of previous finds will indicate the most likely areas to search.

Old maps often provide a key to the site of hidden treasure, and being able to read a map is a decided advantage for any seeker of treasure. Such maps can be found in museums or in old newspapers, and can be photocopied in the course of researching the project. Maps showing the location of old gold mines and gem deposits can be obtained from most state mining authorities. Beware of specially published "treasure" maps, however, for these will rarely lead to anything worth while, and if they do, it is almost certain that someone else has already been to the site. These maps are a source of money only to the people who produce them!

Talking to locals can often help. A farmer who has an old ruin on his property will often know a great deal about it. Since you need his permission to enter his property anyway, it is always worth chatting him up and maybe gleaning a few morsels of information to help locate the possible site of any treasure. As often as not, information that he considers to be superfluous may contain a clue which, when added to previously researched data, will provide an extremely useful lead. Country folk love to talk, and it is surprising how much useful information can be solicited from an aged inhabitant in his local pub. Many a fine treasure has been located for the price of a few beers.

The nature of the site in which treasure may be found can vary enormously. It would be impossible to detail them all in a book such as this, but the following will provide a guide to the more common locations:

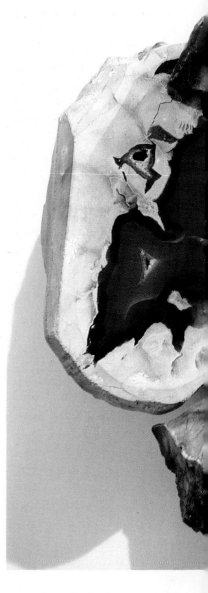

Streams and creeks	Farms and properties	artefacts, old possessions
Abandoned mines	Country auctions	gold, coins, artefacts, possessions
Mullock heaps, dumps, tailings	Country second-hand stores	old possessions
Rivers	Backyards	coins, rings
Caves	School and church fêtes	old possessions
Beaches	gold, gemstones	old utensils, tools, furniture, etc.
Abandoned huts and homesteads	gold, opal	bric-a-brac, furniture, paintings, etc.
Ghost towns	gold, opal	bric-a-brac, old utensils, etc.
Attics	gold, gemstones, coins, artefacts	old lamps, machinery, tools, furniture
Parks	bushrangers' loot, miners' possessions	books, bric-a-brac
Old gardens	wreck artefacts, coins, watches, rings, shells	

4. Recovering the treasure

Having located the site, the search begins. There are numerous ways of discovering and recovering treasure, each dependent on the nature of the treasure and the place in which it lies. It goes without saying that diving gear is essential for underwater wrecks, even in relatively shallow water, while panning dishes are necessary equipment when searching for alluvial gold or gemstones in creeks and streams. Metal detectors are useful, even essential for some types of hidden treasure. The methods used and the equipment required for recovering treasure are described in detail in the section of this book dealing with each specific treasure.

As with all activities, particularly those that involve other people or other people's property, there are a few basic rules which must be observed when treasure hunting. They are fairly obvious and will already be observed by most conscientious treasure seekers, but since failure to follow them can affect and offend others, it is important to mention them here.

Semi-precious gemstones rely for their appeal on their colour and interesting patterns.
PHOTO JOHN CLARKE.

ALWAYS
obtain permission to enter private property. If it is not on Crown land, even the most innocuous creek belongs to someone, and fossicking without first obtaining permission from the owner can lead to unpleasantness. It may even mean the outlawing of all future treasure seekers from that particular site. Few property owners object to treasure hunting on their property providing they are extended the courtesy of first being asked.

ALWAYS
put things back as they were. Nothing is more likely to antagonise a property owner than to find his land pockmarked with holes. Remove any signs of a camp, and restore the working area to its original condition if you want to be allowed back again.

ALWAYS
take care with fires, fences and stock. In bush areas, the need for care with camp fires goes without saying, for a sudden bushfire can cost the property owner dearly. Similarly, ensure that fences are not damaged in any way and that gates are not left open. The loss of a single steer can cost the farmer up to $500, which will not endear treasure hunters to him.

ALWAYS
carry out what you carry in. Never leave rubbish at a camp site, in a creek or at diggings. It is best to carry it out when you leave or, if the farmer is agreeable, ensure that it is *well* buried.

ALWAYS
ensure you are familiar with the regulations when searching in areas such as a national park (not usually permitted), a historic site or a declared shipwreck site. Do not remove *anything* from such sites without first contacting the controlling authority. As a rule, these sites are for *looking only*, and nothing must be disturbed or removed. Heavy penalties are imposed for infringement of the regulations governing these sites.

ALWAYS
ensure that you are prepared for any rigours of weather or environment. When diving on offshore wrecks, ensure that the weather is settled and unlikely to change suddenly, and never dive alone. Similarly, ensure that you have sufficient water and the correct vehicle if your search takes you into desert areas or places far from contact with civilisation. A wise precaution in both cases is to advise the police or a relative of your intentions and your planned date of return.

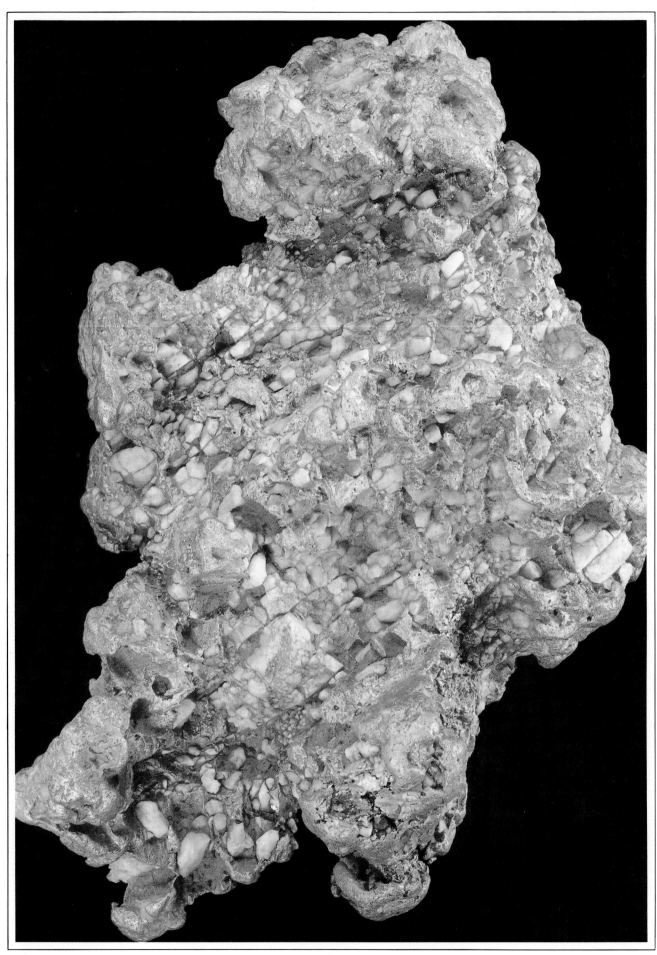

CHAPTER TWO

GOLD
THE GREATEST TREASURE

Gold is not a mineral compound, as is sometimes thought, but a precious metal. A metal that, particularly in recent years, becomes more precious with every passing day. World demand seems never to slacken, not only because of the use of gold in the monetary systems, but also because of its many applications in industry. In Australia today, old workings are being re-examined and new mines opened.

The diggers of the 1850s won most of their gold the hard way — with a pick and shovel or a cradle and pan. With no overheads, and gold fetching a good price, it required only a few grains of quality metal to make the effort worthwhile. Literally thousands of normally sane and conservative Australians walked out of their shops and offices, or abandoned their businesses, their homes, and often their families, to glean easy riches from the glittering streams in the bush. They were joined by thousands more from farther afield — migrants from every corner of the globe, and professional gold-diggers from the worked-out fields of California and Canada.

Today's miners have an uphill battle. Almost all of the easily won gold has gone, and extracting sufficient quantities

Gold in quartz from Temora, NSW.

PHOTO SCOTT CAMERON, COURTESY NSW DEPARTMENT OF MINERAL RESOURCES.

Opposite: The ultimate treasure — this 'Father's Day' nugget was located by an amateur prospector near Hill End, NSW, on 2 November 1979. It weighed 6.2 kg.

PHOTO DAVID BARNES, COURTESY NEW DEPARTMENT OF MINERAL RESOURCES.

to justify full-time mining nowadays requires sophisticated and expensive equipment. Most of the gold recovered is from deep leads or reefs, beyond the reach of amateur prospectors who are limited to surface, or immediate subsurface, mining.

The amount of gold recovered in this way is usually small, and few individual miners make a living from this once-lucrative type of operation. But the lure of the yellow metal works as surely today as it did over a century ago and the prospect of striking a rich lode, or of unearthing a big nugget, attracts hundreds of amateur prospectors every weekend to the gold-bearing valleys and streams. The increasing value of gold makes even small finds worthwhile, and the stimulus created by occasional good discoveries provides the incentive.

If incentive were needed, it was provided in October 1987, when prospector Ray Hall and his gold-seeking family located a reef near Tennant Creek, with potential to exceed any previous single find in Australia's history. The Halls have done well out of gold. In 1986 Ray's daughter Raelene found a nugget weighing 425 grams, while Ray himself recovered gold to a value of around $40,000. Now the new reef, the full value of which will not be known for some time, has assured them of wealth and comfort in their years of retirement. Old prospectors in the area are not surprised; they have known of the rich potential of the area for years. One prospector only recently retrieved over 1000 small nuggets from one area, while another discovered almost 794 grams in a dig only 20 kilometres from the town.

For city dwellers, in particular, fossicking is a rewarding diversion. Even if the pan remains agonisingly empty of glittering yellow specks, a weekend in the open air, deep in the peaceful atmosphere of the bush, pursuing a hobby that can be enjoyed by the entire family, is a reward in itself.

Although a little knowledge is said to be a dangerous thing, in the case of fossicking for gold, it not only helps to avoid pitfalls but can also add considerable enjoyment to the field work. Knowing where to look and what to look for can avoid wasted time fossicking in areas where there is little likelihood of success. Knowing how gold is formed and the type of geological formations in which it is likely to be deposited, can add interest to the hunt. Knowing how to use the fossicking gear can mean the difference between capturing and losing a grain of gold in a panful of murky silt.

The tools are minimal, the delights boundless, and the rewards can be significant.
PHOTO DAVID BARNES, COURTESY NSW DEPARTMENT OF MINERAL RESOURCES.

Right: Reef gold occurs in quartz veins intruded into other rock.
PHOTO DAVID BARNES, COURTESY NSW DEPARTMENT OF MINERAL RESOURCES.

HOW GOLD IS FORMED

Gold is a yellow metallic element which occurs in widely dispersed deposits in the earth's crust. It is rarely found in a pure form, mostly forming a natural alloy with silver, and sometimes with mercury, iron or copper. Gold occurs in many different types of rocks and is found in many different forms, depending on the way in which the rocks were created. Primarily, it seems to be a product of volcanic activity, although the original form and location of the gold can change over long geological periods. There are two principal forms of gold deposits:

Primary deposits are those found in igneous rocks; either in plutonic rocks such as granite, or volcanic rocks such as basalt, where they have been intruded as a result of volcanic activity. The intrusions are known as reefs and are embedded firmly in the host rock, usually in veins, and often in association with a mineral such as quartz. These reefs sometimes reach the surface as the result of crustal folding and movement, but many remain firmly embedded beneath the surface and must be extracted by shaft and tunnel mining.

Secondary deposits. Where gold-bearing rocks outcrop on the surface they are exposed to the processes of weathering and erosion. This action breaks down the rocks and the reef it contains, so that the gold is scattered across the surrounding terrain. Such gold is said to form an "eluvial" deposit. Continued weathering may eventually reduce the rock to a fine detritus, which is then washed into local streams and rivers. Gold, being a heavy metal, quickly sinks to the bottom of the streams, forming what are known as "alluvial" deposits.

Not all alluvial deposits are found in active rivers and streams. Many waterways have, over the years, dried out and filled up. Others became so silted the water could no longer flow and they also dried out and were filled. Some were buried beneath lava flows from active volcanos. In all these cases the layer of alluvial gold becomes buried well below the

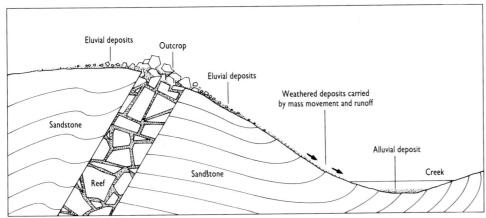

Diagram 1
When a gold reef outcrops above the surface of the surrounding terrain, it is subject to weathering, which breaks it down into gold-bearing detritus. The precious metal scattered across nearby terrain is known as eluvial gold, while that carried by run-off into streams and rivers is known as alluvial gold.

15

An aerial view shows the silt deposits on the inside of the river bends, which have developed into arable river flats. Erosion of the bank on the outside of the bends can be seen by the cliffs at the top of the picture.

The Araluen River in southern New South Wales is a typical gold-bearing river, with heavy deposits of gravel and detritus brought down from its headwaters in the Great Dividing Range.

surface and can only be reached by a mine shaft. Such deposits are known as "deep leads".

"Fools' gold" is the name given to minerals often mistaken for gold. Prominent among these are pyrites and mica. These can be readily identified by the fact that they easily scratch or tarnish and are much lighter than gold.

SECRETS OF THE STREAMS

Since tackling a well-embedded reef is for experienced, well-equipped prospectors, most family fossickers will head downhill to look for the more easily recovered alluvial deposits in nearby streams. With the possible exception of the use of metal detectors, this is the easiest and most certain method of finding gold deposits. Panning a stream has led many a prospector to either a good alluvial deposit, or "colour", which has led to the discovery of a reef upstream.

The secret lies in "reading" the stream, for every waterway has its story and just as a fisherman knows the signs that indicate where the fish are, so the experienced fossicker knows how to read the signs that tell him where to look for gold. It is usually best to head downstream until the fast flow of the mountain creek has slowed to a more gentle run or even to a sluggish meander. Sample panning should be made on the way down, for there are many pockets and runnels, even in fast-moving streams, where grains of gold may have lodged.

Fast-moving water often creates deep pockets, such as those under a waterfall, where grains of gold can become lodged in crevices or among gravel and pebbles. Similarly, obstructions in the stream create back eddies which may suck grains of gold out of the fast-moving stream and into a quiet pocket where they again sink to the bottom. Logs, rocks, sills, or any other feature that creates a pocket of slow-moving water, can be a potential hiding place for the heavy grains of gold.

But it is farther downstream, when the hectic flow of water has slowed, that the discovery of gold deposits is more predictable. Ships' navigators have for long known that the water on the outside of a river bend is deeper than on the inside, mainly because the faster rate of flow around the outside bank scours a deep channel. The slower water on the inside deposits silt which builds up into a shoal bank, sometimes surfacing to form a sand or gravel spit. Since gold sinks to the bottom in slow-moving water, these banks are prime spots to pan for gold.

Farther downstream again, the river will meander through lowland flats. When the river floods, some of the bends are cut off to form oxbow lakes. Deposits of silt, often containing gold, build up at each end of these lakes as the river is deflected into its new course. Later again, as the river moves towards its mouth, it spreads out into a delta, and the slower flow of the water causes silt and minerals, often including gold, to be deposited in what are known as alluvial fans.

THE ART OF FOSSICKING

There are many different ways of extracting gold from the pay-dirt in which it is lodged, the most common involving the use of a panning dish or a metal detector. More sophisticated equipment, such as cradles or dredges, is usually too expensive and impractical for family fossicking. In areas where there is no water, a dry blower will be necessary, and if a primary deposit is to be tackled, then picks, crowbars and other heavy gear will be necessary.

But generally speaking, a family fossicking for gold will need only minimal equipment, the cost of which can range from a few dollars to a few hundred. Panning dishes cost about $15, so even with one for each member of the family, the bill is not going to ruin the holiday. Metal

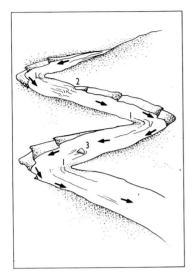

Diagram 2
Reading the stream. The main flow of the stream sweeps fast around the outside of bends. The slower flow on the inside, as well as behind obstructions and beneath waterfalls, allows the relatively heavy gold to drop to the bottom. Gold will be deposited at
1. point bars on the insides of curves and bends;
2. waterfalls and areas where flow is disturbed;
3. behind obstructions.

detectors, by contrast, can cost several hundred dollars and make a fossicking excursion a fairly expensive exercise.

Much will depend on the experience, the dedication and the ambitions of the fossickers. A trip to an old mining site with no stream, or with only limited water, will not produce much for those armed with panning dishes, so a metal detector is essential to achieve any worthwhile results. To minimise costs, it might be wise to hire a detector for the first couple of trips. Once the bug has bitten and the family become avid fossickers, then purchasing a new detector will make sense.

If, on the other hand, the chosen site is a stream or river, panning dishes will provide plenty of action as everyone learns the techniques of swirling the gold from its watery bed. Most metal detectors can be used in water, so it might be worth combining both types of fossicking to determine which will be most suitable for future excursions. Either way, the initial trips should not entail the purchase of expensive equipment. As a rough guide, the items below should be considered as basic for any fossicking trip.

The basic equipment for extracting gold (or gemstones) from washdirt consists of two or more sieves with varying grades of mesh and a panning dish equipped with riffles. Cost in 1988 would be about $30.

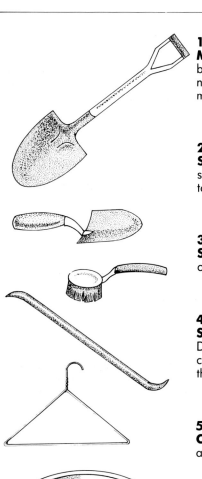

1.
Miner's shovel. Preferably with a long handle to reduce the amount of backbending required. This should be a light or bantam shovel with a pointed nose to get into awkward corners. Big shovels handle a bigger volume but are much heavier to carry.

2.
Small trowel. This is useful for getting into corners that are too tight for the shovel. Grains of gold often lodge in such corners and the trowel will be necessary to get them out.

3.
Small brush. Preferably with fairly stiff bristles. This is also used to get pay-dirt out of awkward corners and crevices.

4.
Small jemmy. Necessary for prising out rocks or other obstructing material. Depending on the nature of the terrain, a geologist's hammer, a miner's pick or a crowbar might be more suited for this work. However, these are mostly heavier than a jemmy and more cumbersome to carry.

5.
Coathanger wire. When all else fails, this will probe into the tightest crevices and corners to dig out reluctant pay-dirt. It is also useful for extracting wash-dirt.

6.
Panning dish. Special dishes made for gold panning can be obtained from prospectors' stores. They are fitted with a "riffle" or groove to catch fine grains of gold as they are sifted from the sediment. Traditionally, panning dishes are made of tin, but modern plastic dishes are light, easy to use and very durable.

Less orthodox methods of recovering alluvial gold include sluicing it from the silt of river beds. Even small portable dredges like this, however, are prohibited in some areas because of the damage they can cause.

PHOTO BOB GREIG, MINERS' DEN.

7.
Container. Some fossickers prefer to pan only enough to concentrate the wash-dirt. They then save it and extract the gold later, processing under more controlled conditions. In this case, a billy, or similar container — known as a "poverty pot" — will be necessary. For those who pan out the gold on the spot, a small coffee jar is ideal for carrying the grains or nuggets.

8.
Tweezers. A pair of eyebrow tweezers can be very handy for picking out tiny grains of gold from the sediment in the panning dish.

9.
Magnifying glass. Useful for detecting small flakes or grains of gold "colour" among the wash-dirt.

10.
Sieve. The need for a sieve will depend on the wash-dirt being panned. Where large pebbles or rocks, twigs and other debris are scattered among the detritus, a sieve will remove them and make panning easier.

11.
First aid kit. Essential in any form of fossicking.

12.
Maps. Geological, topographical and mining maps are useful in determining the best spots to fossick.

USING THE PANNING DISH

A description of how to "read" a stream or river for the most likely fossicking spots was given earlier. Having located such a spot and pitched camp, the real work begins. The panning dish must be dark in colour in order to show up specks of gold, so a new metal dish must be blackened over a fire. Most modern plastic dishes are dark in colour and ready to use without further treatment.

A close look at the river sediment should indicate a good spot from which to take the wash-dirt. Crevices and pockets created by swirling water, tree roots or fallen logs which have trapped sediment, or areas near the junction of a stream and the river are typical of the places where gold-bearing pay-dirt can be found. Assuming that the sediment does not need to be sieved, the panning procedure should be as follows:

1. Shovel wash-dirt into the dish until it is about two-thirds full. Hold the dish in both hands, with your fingers beneath the rim and thumbs on top, and immerse the dish in the stream until it is just beneath the surface.

2. Swirl the dish so that the wash-dirt moves in a circular pattern around and across the inside of the dish. If the sediment is mostly heavy clay, it may be necessary to break it up.

3. Tilt the dish forward and raise the back out of the water, keeping the front lip just level with the surface. Continue the swirling, rotary action until the water becomes clearer, then gently tilt the dish from side to side.

4. Rotate the dish again, and while the water is swirling around inside, tilt it forward until the water begins to run over the edge. The coarser gravel will move to the lip of the dish and can be edged out with your thumb or carefully tipped out with the surplus water.

5. Continue the process, reducing the amount of water and wash-dirt in the dish, taking care not to let any fine sediment escape, as this may contain flakes of gold.

6. When the water is well down the dish and the wash-dirt is becoming thick, repeat the procedure, immersing the dish again and panning away the surplus water and coarser sediment. It is important always to keep sufficient water in the dish to keep the wash-dirt loose, or the gold will not be able to sink through it.

7. Now place the lower lip of the pan into the water and gently scoop some water into it, swirling at the same time. As the wash-dirt concentrates in the pan, tilt it more frequently to remove the coarser material, as described above, but with even greater care. This procedure is continued until the wash-dirt has reduced to about a spoonful and any grains of gold should become visible. Remove them by pressing a wet finger onto them and transferring them to the gold jar.

8. With only a small quantity of water, swirl the remaining dirt around the corner of the pan; then, with a steady flick, remove the water to the other side. The dirt, in attempting to follow the water, will spread across the bottom or catch in the riffle, thus giving you an opportunity to see any further specks or examine it with a magnifying glass.

Using the panning dish
1. The water does the work initially, flushing away soluble clay and sediment.
2. Larger rocks and stones can be discarded after examination.
3. As the washdirt is concentrated, closer examination is necessary.
4. Colour! The reward for painstaking work. Even small pieces of gold soon add up to worthwhile treasure at today's inflated prices.

PHOTO DAVID BARNES, COURTESY NSW DEPARTMENT OF MINERAL RESOURCES.

As with every craft, learning the skills of panning comes only with practice. Experts can swirl the dish with apparent ease, yet without letting even the finest speck of gold escape. Beginners will fumble and bumble and lose frustratingly large amounts of wash-dirt before finally acquiring the knack and achieving moderate success.

There are many different techniques, some more suited to certain types of sediment than others. The description given above is a basic routine which can be adapted to suit the texture of the wash-dirt, the flow of the water, or any other feature that is unique to a specific area.

The secret is in removing the finer wash-dirt without losing any gold it may contain, and this is where only practice can make perfect. Beginners should tackle only the first few steps until they become familiar with removing the coarse gravel, leaving the finer washes to those with more expertise. As they become familiar with the handling of the dish they can progress to further steps until, having mastered an art that is as much a part of Australia as gold itself, they can swirl right down to the final wash and enjoy the adrenalin-pumping excitement of spotting the glittering metal.

Not all alluvial gold is found on the surface in streams and rivers. Old river beds which contain quantities of gold often become buried with time, either as a result of the stream drying out and filling up with dust and soil, or as a result of some geological action in the past. Landslides in mountainous country frequently change the course of a stream, and the meandering waterways of the lowlands, where often the best gold deposits are found, tend to change shape and direction with floods or droughts, leaving stretches covered with deep layers of silt.

Volcanoes of earlier geological periods frequently covered river beds with streams of lava which buried the alluvial gold metres deep beneath what is now basalt. The majority of deep leads in the eastern states are covered with layers of basalt. To reach these leads it is necessary to dig a shaft, and the depth may vary according to the nature of the overburden and the time the lead has been buried. In areas such as river flats, where the layer of gold has been covered with continuous silt layers from seasonal floods, it may be only a metre or so below the surface. For this reason, it is worth digging a few test holes when fossicking river flats or deltas. A sample of sediment shovelled out from a metre below the surface may show signs of gold, where none was found in the surface sediment.

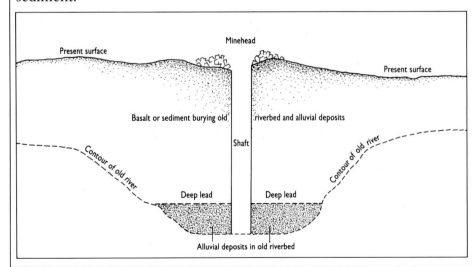

Diagram 3
Old riverbeds are often buried beneath the basalt of an ancient lava flow, or dry out and are covered by a build-up of sediment. Alluvial deposits then become deep leads, and access is usually only possible by shaft mining. Shafts can be several hundred metres deep.

Since the river would, in its earliest stages, have run over a rock bottom, the excavation should, if possible, be sunk to bedrock so that the deepest alluvial deposits can be checked. Numerous amateur fossickers may have worked the stream and its surface sediments, but few will have taken the trouble to test the buried layers. Taking the hole down to bedrock ensures that you are able to check all the deposited layers of the river, from its original bed to its current level. If a number of such test holes in one area do not show at least a trace of gold colour, then it is time to move on.

Deep leads require a properly constructed shaft to be sunk and this is beyond the scope of the average family fossicker. Because of the dangers inherent not only in sinking the shaft, but also in mining the lead at the bottom, this kind of gold mining is for very experienced prospectors only. A little knowledge can well be a dangerous thing in this context, and a book such as this could not provide the information required to mine deep leads safely. Only experience in working such mines with professional miners will provide the necessary knowledge and reduce the risks of possible accident.

DRY BLOWING

In the more arid parts of the country, where rivers and streams do not exist, there is nothing to carry away the detritus when a reef is exposed and weathered. The broken gold, often contained in quartz, is scattered across the surface and not concentrated into leads, as it would be if it were washed into a stream or river to form an alluvial deposit. In some dry areas, where extinct rivers once concentrated the weathered gold, deep leads may be found buried beneath the arid surface soils, but many inland areas have no concentration of the precious metal other than in the reefs that have not as yet weathered.

Two main problems are encountered when fossicking in such areas — the scattered nature of the deposits, requiring a great deal of effort to sift the gold from the dirt, and the lack of water for panning. In these areas, water is either not available at all or is too precious to waste on panning, so a method of recovering gold from the dirt without water is used. Known as dry blowing, it requires the use of a special piece of equipment called a dry blower.

There are many different forms of dry blower, many constructed by prospectors to suit a particular need. Basically, it consists of an open framework with a number of different levels. Pay-dirt is shovelled into the top level where a sieve removes large rocks, stones and other debris. It also catches any sizeable nuggets which might be concealed in the dirt. As the residue from the sieving drops through to the next level, a set of bellows, or a power-operated blower, is used to blow the dust away. Each level has a finer screen than the one above, so that gradually the detritus is reduced to a fine sand or gravel. This residue is then transferred to a dish and the blowing process repeated very carefully by mouth.

In gradual stages, just as with panning in water, the dirt and gold are separated and any grains or specks of gold are removed. Many prospectors use two dishes, transferring the dirt from one to the other, blowing away the dust each time. The heavier metal will separate gradually from the dirt, although not as easily as when sluiced in water.

The most basic, and probably most arduous, form of dry blowing — by mouth — demonstrated by an old-time miner in a dry creek bed in Western Australia.

Dry blowing requires more patience and persistence than panning, and since it is usually undertaken in hot, dry conditions, it is not an over-popular method of prospecting among amateur prospectors!

USING METAL DETECTORS

Probably no single factor has affected treasure hunting as much as the introduction of metal detectors. Like so many devices that ultimately become useful in normal life, the metal detector was first developed as a weapon of war and is still used extensively for that purpose. It provides an effective means of locating mines, booby traps and other explosive devices that have been concealed, usually under the surface of the soil. During World War II, trained technicians, known as sappers, used what were then called mine detectors to sweep a path through minefields designed to stop the advance of tanks, vehicles and even foot soldiers.

Away from wars and conflicts there is happily a much more pleasant use for these electronic devices — locating metallic treasure. Probing beneath the ground with their low-frequency oscillations, metal detectors can, depending on their construction and sensitivity, locate ferrous or non-ferrous objects for some surprising depth. The two most popular uses are for locating gold nuggets and for searching beaches, parks and old buildings for lost coins and other personal treasures.

The metal detector has revolutionised gold fossicking to the extent that many of the long-abandoned mines and workings are now being re-examined by prospectors with metal detectors. Whereas, with previous methods, the pay-dirt had to be laboriously sifted, washed and examined by hand, now the detector takes away much of the physical effort. By careful tuning and use, no more effort is required than to sweep the loop, or head, of the detector over potential gold-bearing ground, and listen carefully for any change in the audio signal.

Apart from the ease of use, there are many other advantages with this equipment. Areas that once were difficult to prospect, such as outcrops of rock, will often respond to the oscillations of the metal detector. Ray Hall's discovery of a major gold-bearing reef to the south of Tennant Creek owed much to the use of a detector. The happy prospector and his family used a metal detector to examine each new rock face they opened, and ultimately their perseverance paid off when the detector indicated a big gold lode. The million-dollar "Hand of Faith" nugget, which weighed 27.2 kilograms, was discovered in Victoria in 1980 by an amateur using a metal detector.

Indeed, it would be safe to say that no serious prospector, amateur or otherwise, would set out to look for gold without an electronic detector. While panning is still considered by most to be the best way to get gold from a creek bed, waterproof metal detectors are impinging even on this last bastion of old-world prospecting. Certainly the clean, easily used device is more pleasant and more rewarding than the dusty dry blower in areas where there is no water.

Treasure hunters who look for lost treasure other than gold also have reason to be thankful for the invention of the metal detector. Some prospectors spend lucrative weekends "walking" their detectors over beaches or parks where people have lost coins, watches, rings and other valuables.

Searching the surrounds of a derelict building can unearth all kinds of

A metal detector being used on an abandoned puddling site.
PHOTO BOB GREIG, MINERS' DEN.

treasures, some with monetary value, some with historic value. Museum expeditions looking at old sites often find invaluable relics which, because they were buried randomly across the site, might never have come to light but for an electronic detector.

There are a number of different types of metal detectors on the market, some more sophisticated than others, some designed for a specific use. Without going into great detail about these instruments, it is worth examining the basic features of those most commonly used for treasure hunting. Like all equipment, metal detectors only give of their best when correctly tuned and handled, and novices would be well advised either to take a course in the use of such equipment or obtain assistance from an expert before attempting to buy and use one.

Many fossicking and treasure-hunting clubs and organisations provide instruction and advice to members and also organise field days in which new equipment can be tuned and checked under the eye of experienced users. Club members are renowned for their willingness to assist new-comers in any field, and joining a fossicking or treasure hunters' club would be a good move for any beginner. Apart from advice and assistance in purchasing a metal detector, it is always useful to be able to watch how experienced operators go about using the equipment. Picking up tips from the experts can avoid many hours of fruitless effort and considerable disappointment.

The two detectors most commonly used by Australian treasure hunters are the VLF/TR and the VLF Discriminator. The initials VLF, which are common to both, refer to the Very low Frequency of the wavelength used in the detector. Very low frequency wavelengths have the ability to penetrate deeper than high-frequency wavelengths. The storm about siting one of the world-spanning Omega navigation stations in Australia related to its very low-frequency transmissions. Unlike transmissions on higher wavelengths, the Omega wave can penetrate up to 30 metres below the surface of the oceans, thus enabling submarines to operate navigational equipment without the need to surface.

The VLF Discriminator is the most popular detector for novice treasure hunters. As its name implies, it can discriminate between ferrous, non-ferrous and mineral objects while at the same time reducing or cancelling the effect of minerals in the ground. The usual frequencies available with this instrument are between 3 kilohertz (kHz) and 15 kHz.

All metal detectors operate on batteries and most have dials or meters in addition to headphones. While the meter will indicate response from a buried object, the audio signal in the headphones often makes it easier to detect weak signals from small objects or those that are buried deep beneath the surface.

Tuning the detector

Like a musical instrument, the metal detector must be tuned before use. While tuning is basically the same for all instruments, some brands require slight adjustments to the basic routine, so it is important to follow the manufacturer's instructions until familiar with the tuning procedure. Tuning involves two main steps — setting the threshold level and setting the ground-cancelling control.

Setting the threshold level is done by first switching on the instrument and setting all controls as instructed, then allowing the instrument to

Metal detectors have revolutionised fossicking, particularly for gold. Accurate tuning for the terrain in which it is being used can mean the difference between locating a nugget and just missing it.

PHOTO BOB GREIG, MINERS' DEN.

warm up. A constant tone will be heard, and this is the tone to be adjusted. The ground beneath the loop of the detector must be clear of any metal objects or a false threshold will be set. The loop is then raised about 60 centimetres above the ground and the tone reduced to a low, but still audible, level. This is known as the threshold level.

The loop is then lowered to about 25 millimetres above the ground and the memory switch released. If the threshold tone changes, the ground-cancelling control requires adjusting, so the loop is raised again, and the ground-cancelling control adjusted. If, on lowering, there is still a change in the threshold tone, the procedure must be repeated. Only when the loop is lowered to within 25 millimetres of the ground without any distinct change in the level of the tone, is the instrument ready for use.

As the detector is moved over the ground in use, the level of mineralisation beneath it may change, in which case adjustment of the ground-cancelling control may be necessary to retain the threshold level of the tone. Some instruments are equipped with auto-tuners which automatically adjust the ground-cancelling control as the threshold changes. This can save a lot of irritating adjustment and readjustment when the detector is used over ground with varying mineral content.

Using the detector

Although most modern metal detectors are not heavy, anything can become heavy when it is held for a long time. Some types of detectors have the main body of the instrument separate from the stem and loop, so that it can be supported by a strap over the shoulders. Probably the most important aspect of using the detector, and one which reduces the stress on arms and body, is to take the correct stance.

The length of the stem is usually adjustable, so it can be lengthened or shortened to ensure that when it is located at the correct height above the ground, it is not causing the operator to stand awkwardly. This is another good reason for joining a club or group, for much can be learned from watching experienced operators, and a great deal of shoulder or arm ache avoided. Practice will soon make perfect, but practice with experts can reduce the time taken to achieve perfection.

Detecting is done with a sweeping or swinging action. The loop should be held close to, and always parallel to, the ground. Where the ground is heavy with minerals, the loop should be about 25 millimetres above the surface. Where there is little or no mineralisation, the loop can be lowered. Indeed, it can be operated while actually touching the ground, although this can cause wear or damage to the base of the loop and is not advisable unless the loop is equipped with scuff pads.

Sweeping is carried out by swinging the shoulders with an easy motion from side to side, allowing the arm to swing on a little after the shoulders have stopped. This may sound complex but it is an easy and natural action. The more natural the swing, the easier the sweeping will become and the less the strain on the body. As with all new actions or movements, practice soon develops a technique which synchronises all muscles and limbs involved in the movement, thereby reducing the strain on any one area.

In order to ensure complete coverage of the ground being swept, a system of lanes should be planned. Each sweep should overlap the previous sweep and each lane must overlap the previous lane. The width of the scan is widest on the surface, becoming narrower with depth until it comes to a point at its maximum penetration. This means the deeper the target, the less chance of detecting it. Overlapping both the sweep and the lanes ensures the best coverage even for deeply buried objects.

As mentioned at the beginning of this section, only the basic outlines of using a metal detector are provided here. Practice and study are necessary to perfect the techniques that will produce the best results. There are numerous books on the subject and many treasure-hunting clubs where information can be gleaned. Beginners would be well advised to make use of all possible sources of information before buying, let alone using, a metal detector.

Many metal detectors can be used in water and are therefore useful in wide sedimentary deposits such as this braided section of a stream bed, where a panning dish could only be used to examine one small area at a time.

CHAPTER THREE

THE GOLDEN STATES

NEW SOUTH WALES

The old goldfields of New South Wales are so extensive and so widely scattered that it is impossible to list every fossicking site that might yield gold. Despite this, New South Wales is not a prime gold-producing state. Since 1851, the state has accounted for only 8.5 per cent of the total gold production in Australia, falling well behind Victoria, Queensland and Western Australia. Today, almost all of the country's gold comes from Western Australia, with no major producers among the eastern states.

Almost all the gold recovered in New South Wales is the result of fossicking by amateur treasure hunters. The only commercial output comes as a by-product of base metal sulphide concentrates in treatment plants at Broken Hill and Cobar. This means that all the old goldfields are available as hunting grounds for fossickers. Since one of the most productive areas of gold recovery is among the workings and tailings of old mines, virtually all the gold-bearing regions of New South Wales are at the fossicker's disposal.

There are two main gold-bearing belts,

Map 1
Most of the known gold deposits in New South Wales fall within two fold belts, which are separated by a "corridor" running north-west from around Sydney.

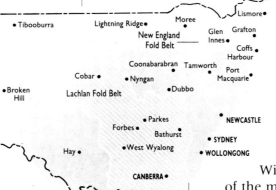

Opposite: Where it all began. The first payable gold in Australia was found at this spot, where Summer Hill and Lewis Ponds creeks flow together. It was named Ophir, after the Biblical city of gold, and still produces colour to this day.

so extensive that between them they cover almost half the state's surface area and range from the coastline inland to Bourke and Tibooburra. The Lachlan Fold Belt, which is the larger of the two, takes in the south coast from around Batemans Bay to the Victorian border and covers a wide swathe of countryside running in a north-westerly direction, terminating roughly on a line between Broken Hill and Bourke.

The New England Fold Belt covers the coastline from Newcastle to Coffs Harbour and again runs in a north-westerly direction, terminating roughly on a line drawn between Tooloom and Narrabri.

Within these two belts lie most of the major gold deposits in New South Wales, the only other deposit of any significance being in the Tibooburra area.

The geological structure of the state indicates that many major gold deposits may still lie awaiting discovery. An important deposit was found in recent years at Gidginbung, near Temora, and further exploration in the area revealed a number of other promising prospects in the belt extending from West Wyalong to Adelong.

The prime factor common to most significant gold finds appears to be their association with past volcanic activity. Since there are numerous regions in New South Wales which meet this criterion, geologists are optimistic about future gold finds.

Although new gold discoveries are liable to be in the form of deep reefs and veins requiring extensive mining operations to recover the metal, the presence of reefs indicates the possibility of alluvial deposits in nearby streams and creeks. Similarly, many major reefs have been traced after the discovery of "colour" in nearby creeks.

The presence of gold in any form means there is potential for the enthusiastic fossicker in that location. Where it is not found in local streams and creeks, it may be found in the mullock heaps and tailings of the mines, or in adjacent ground not previously worked.

The fact that an area has been worked before does not mean there is no gold left to find. The early miners often missed smaller deposits and, in any case, weathering and eroding of the surrounding land by heavy rainfall soon replenishes deposits in creek beds. Even areas well-worked by Chinese miners, such as the Palmer River, are producing modest finds again. Mullock heaps that have lain unattended near old shafts or abandoned workings for the past 130 years can produce surprising returns for assiduous fossickers.

The price of gold is now such that even commercial mining organisations are reworking old sites. The amount of the precious metal required to make a reasonable profit is far less than was required when the goldfields were first worked, so reworking an old site can be a profitable exercise, particularly for amateurs with low overheads.

The goldfields described here have been selected because they offer the best potential for amateur fossickers, taking into account not only the potential for finding gold, but also the ease of access and the needs and requirements of family fossickers.

The numerous abandoned mine sites in New South Wales are well worth working, particularly with a metal detector.

Bathurst district (north)

The first payable gold in Australia was discovered in this area, which remains to this day one of the most rewarding for amateur fossickers. Centred on Bathurst, a rich band of gold-bearing country extends northwards to the vicinity of Gulgong and southwards almost to Crookwell, across a wide swathe of the western slopes.

The most popular fossicking areas to the north of Bathurst are in the Hill End—Sofala area where not only was gold found in quantity during the latter half of the last century, but where some of the largest nuggets were unearthed. The Holtermann nugget, which weighed 235,143 grams and had a gold content of around 93,300 grams, was discovered in 1872 at Hill End, and the Kerr Hundredweight, weighing 74,648 grams, with a gold content of 39,564 grams, was unearthed nearby in the Turon River. Good gold is still panned from the creeks and rivers in this area, including the Turon River itself.

Hill End, a fine old mining town, preserved under a Historic Site declaration since 1967, is a good starting point from which to work the streams and creeks of the area. An easy 86 kilometre drive from Bathurst, mostly along sealed roads, the town has retained the atmosphere of the gold-rush days in its restored buildings. Nearby Sofala, on the Turon River, has produced good shows of gold in recent years, while 33 kilo-

Map 2
Major gold-bearing regions to the
north of Bathurst.

metres to the north, the creeks and streams near Hargraves are a fossicker's paradise.

The Cudgegong River, which is a tributary of the mighty Macquarie River, has been a source of gold since the early days and is still popular with fossickers. It can be reached from Mudgee or Gulgong, both of which have featured prominently in past gold rushes and still produce small shows in local streams and creeks.

The Macquarie River must be considered a fine carrier of alluvial gold, for much the same situation exists to the north-west of Bathurst, on the western side of the river. Many of the creeks and streams which feed into the Macquarie from this side are renowned for their gold potential. Stuart Town, some 60 kilometres to the north of Orange, was the scene of a huge gold rush in the latter part of the nineteenth century, and an estimated $5.5 million worth of gold was recovered before 1914, when commercial operations ceased.

The old railway station at Stuart Town, once one of the most lucrative of western goldfields, and still offering good pickings for amateur fossickers.

Stuart Town, which was once called Ironbark and is the subject of Banjo Paterson's well-known poem, "The Man from Ironbark", is little changed from those hectic days, with mullock heaps still prominent beside the main street. Gold is still recovered from creeks in the area and access is easy along a sealed road from Orange or Wellington.

Also on the west side of the Macquarie River, on one of its tributaries, is the spot where Edward Hargraves found the gold that put Australia on the world map. He called the spot Ophir, after the biblical city of gold, and to this day fossickers still recover small amounts of gold from the area. Ophir is within easy reach of Bathurst along a moderate dirt road that leaves the Mitchell Highway at Orange. It is popular not only for the gold that might be found there but because it retains much of the atmosphere of Australia's first gold rush. Although nothing is left of the original settlement, shafts, tunnels and mullock heaps surround the creek junction.

Bathurst district (south)

Since it is still relatively undeveloped, the countryside to the south of Bathurst is a popular area with more experienced fossickers. The atmosphere of the old gold-rush days is alive in the hills and steep gullies, where rushing creeks have all the classic appearances of gold-bearing streams, and the winding dirt roads that were originally built to transport gold. The few small towns and villages in the area are, in many cases, unchanged from those roaring days when fortunes were made in the surrounding hills and the creeks. This was an area where commercial mining was less prominent and most of the gold was recovered by individual prospectors.

Map 3
Major gold-bearing regions to the
south of Bathurst.

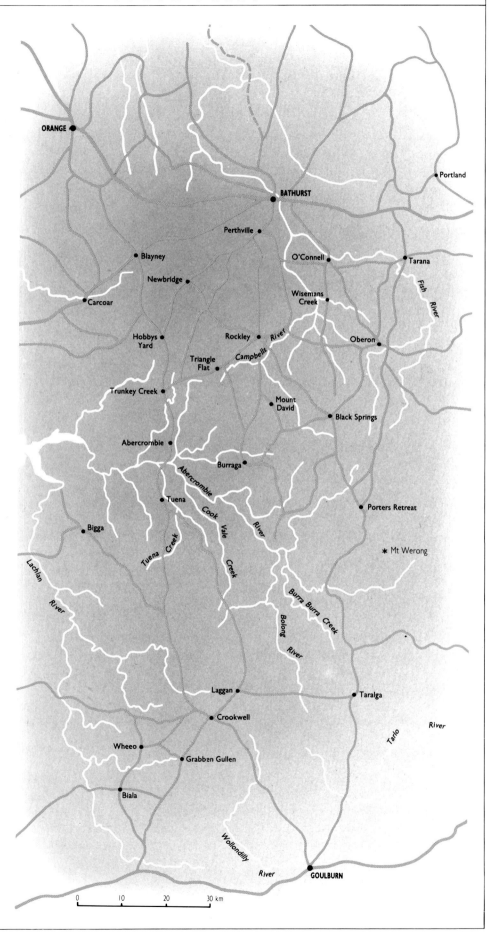

ORANGE

Portland

BATHURST

Perthville

Blayney

O'Connell

Tarana

Newbridge

Wisemans
Creek

Carcoar

Rockley

River

Oberon

Campbells

Hobbys
Yard

Triangle
Flat

Trunkey Creek

Mount
David

Black Springs

Abercrombie

Burraga

Abercrombie

Porters Retreat

Tuena

Cook Vale

River

* Mt Werong

Bigga

Tuena Creek

Creek

Lachlan

Burra Burra Creek

River

Bolong

River

Laggan

Taralga

Crookwell

Tarlo

River

Wheeo

Grabben Gullen

Biala

Wollondilly

River

GOULBURN

0   10   20   30 km

The gold-bearing district lies in the triangle formed by lines joining Bathurst to Mount Werong and Tuena. The only town of any note is Oberon, which sits on the fringe of the triangle and is the best spot from which to launch fossicking forays. Rockley is a historic village with links to the golden days, as are Trunkey Creek, Tuena and Black Springs, but these have little to offer fossicking families other than adjacent camping and a few facilities at the village store. Like the rest of the region, most are accessible mainly over dirt roads, though normal two-wheel-drive vehicles can cope quite easily with all but off-road tracks.

The main potential for gold in this area lies in the rivers near Foleys Creek and Black Springs (the Fish and Campbell rivers), Abercrombie (Abercrombie and Isabella rivers and Rocky Bridge Creek), Burraga (Isabella River), Tuena (Tuena Creek), Porters Retreat (Retreat River), and in Native Dog Creek, between Oberon and Rockley. The streams that feed these rivers all have good potential and some will also produce gemstones, including sapphires. Most sites have camping areas in or near a good fossicking spot, but Oberon or Blayney are the only towns with good facilities. Access to the general area is possible from Goulburn (via Crookwell), from Bathurst, Oberon or Blayney.

Although it is possible to fossick anywhere in New South Wales, other than in national parks and on mining leases and private property (without the consent of the owner), specified fossicking areas are listed by the Department of Mineral Resources. Maps of these areas are available on application to the department in Sydney.

Native Dog Creek is one of many creeks in the area that are rich with gold and gemstones.

Adelong district

This is one of the richest gold districts in the state, the first discovery being made on the top of Victoria Hill in 1857. This was reef gold, but as is so often the case, alluvial gold, which was eroded from the reefs and carried into the valleys by run-off, occurs in most of the streams nearby. The prominent locations are Adelong Creek and Golden Gully, on the eastern flank of Victoria Hill. Gold has been found along a 26 kilometre length of Adelong Creek from about 2 kilometres downstream from the town of Adelong, to the creek's junction with the Murrumbidgee River.

This part of the Adelong Creek was worked mostly with dredges and sluices in its heyday around the turn of the century, when an estimated total of 13,000 kilograms of alluvial gold was recovered. Depleted stream deposits can often be replaced by rains over a period of years and, in any case, a certain amount of gold was always missed by the dredges, so it is always possible that worthwhile deposits could lie in the sediment of this creek. Good finds have been made by amateur fossickers panning the detritus in this stream and in the immediate section of the Murrumbidgee River into which it empties.

Access to the area is easy via the Snowy Mountains Highway from Tumut or the Hume Highway, and the township of Adelong is well organised for visiting fossickers with a historic hotel, a motel and caravan park. Little more than an hour's drive along the Snowy Mountains Highway is another famous gold town, Kiandra, while to the north and south are Gundagai and Tumbarumba, respectively. All these areas have gold potential in their creeks and all make good fossicking grounds.

The belt of gold deposits that covers these regions on the fringe of the Snowy Mountains is believed to extend north to Temora and West

The bleak, storm-swept highlands of the Snowy Mountains region have produced many rich goldfields in the past.

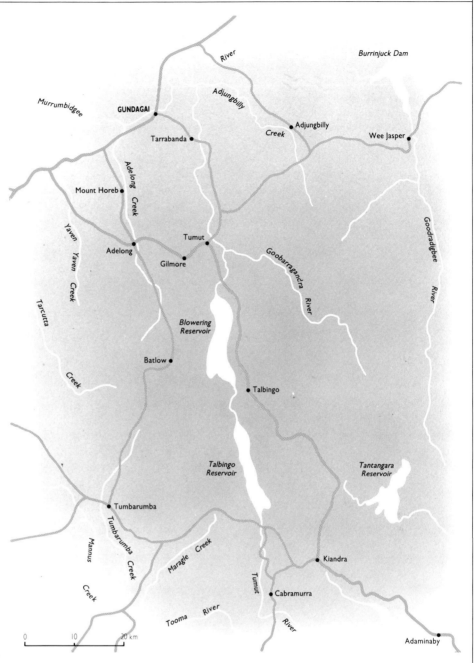

Map 4
The gold-bearing region around
Adelong.

Wyalong and also to Young, where alluvial deposits have been found in profusion. Recent exploration has revealed an important deposit at Gidginbung near Temora. Prompted by this discovery, further exploration of the belt between West Wyalong and Adelong indicates good prospects for future gold mining in this district. For this reason, fossickers would be well advised to work lesser-known creeks in the area rather than the accepted gold-bearing streams. In country with this potential, there is always the chance of striking a new deposit and hitting the proverbial jackpot.

New England Fold Belt
Although better known for its sapphire deposits, the New England region now has the largest number of gold mines in New South Wales. While

Above left: Disused mines are scattered across the entire New England Fold Belt. This is an old tin mine near Copeton.
PHOTO PAUL HANKE.

Above right: A creek near Barraba, one of the most prolific of the New England Fold Belt's gold producing areas.

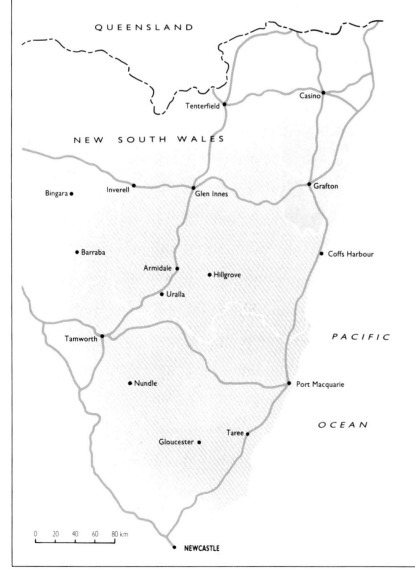

QUEENSLAND

Tenterfield

Casino

NEW SOUTH WALES

Bingara

Inverell

Glen Innes

Grafton

Barraba

Coffs Harbour

Armidale

Hillgrove

Uralla

Tamworth

PACIFIC

Nundle

Port Macquarie

OCEAN

Gloucester

Taree

0   20   40   60   80 km

NEWCASTLE

Map 5
The New England Fold Belt.

most of these are fairly small-scale operations, it does indicate the potential for gold in the streams, particularly in the vicinity of Nundle and Uralla. Working the creeks and streams in this region has a pleasant second string — if you don't find gold, you might find sapphires!

The southernmost area known to contain gold deposits is in the rugged mountain terrain near Barrington Tops, particularly to the east of Scone, in the vicinity of Stewarts Brook and Moonan Brook, where a large nugget was found about ten years ago. On the western side, a popular fossicking area is around the small township of Nundle, particularly in Opossum Creek. Access roads to the streams in these areas can be rough and may require a four-wheel-drive vehicle. Since most of the creeks are on private property, permission must be obtained from the owners before entering.

North of Tamworth, the belt of known gold deposits divides and forms a "V", with a north-westerly arm extending to Bingara, and the north-easterly arm stretching through Uralla and along the eastern side of the Great Dividing Range to Tooloom, then across the Queensland border to the west of Warwick. The deposits in these northern areas are not as concentrated as in the Lachlan Fold Belt, but good finds have been and can still be made in the mountain streams.

Hillgrove, to the east of Armidale, was one of the state's major goldfields, producing over 15,000 kilograms of gold. Undoubtedly, good pickings still remain for the enthusiastic amateur in Bakers Creek, on which the huge mine stood, and other streams in the area.

Gold is still mined in places around Uralla where, again, sapphires are often recovered from the same streams. Glen Innes, north along the ranges, is the location of one of the major sapphire fields in Australia, so fossicking in this area should be rewarding in one way or another.

Considerable gold was taken from the Upper Clarence River Valley, particularly in the areas around Tabulam and Tooloom. The "Big River" draws its water from thousands of creeks and streams in the eastern escarpment of the Great Dividing Range, any one of which could reward energetic fossickers. Since much of this area is well removed from civilisation and access is difficult, fewer prospectors will have worked the streams and the chances of a find must be good.

The hills and valleys of the rivers on the western side of the ranges from Bingara, south past Barraba, are well-known mining centres, with many active mines in operation to this day, some still recovering gold. The Gwydir River and in particular the Bobby Whitlow and Gouron Gouron creeks, have produced good showings of gold, and the belt seems to run southwards, passing to the east of Barraba. Once again, the terrain in these areas is fairly rugged with poor access which, although making a fossicker's life difficult, enhances the prospects of finding a new deposit of gold or gemstones.

Because most of the gold deposits in the New England Fold Belt lie across the rugged highlands of the Great Dividing Range, fossicking is generally far more rugged than in many areas to the south. People who fossick in these areas should be experienced and well equipped, and the vehicles used should be of the right type.

Apart from the difficult terrain, climatic conditions can be severe in the highlands. Snow is common in winter and extremely low temperatures are the norm. Even in summer, rain induced from the maritime winds in these high altitudes can cause temperatures to drop considerably, creating hazardous mountain conditions and making roads impassable. Fossicking parties without bush and mountain experience should confine themselves to the streams and creeks within reach of the main centres such as Uralla, which has numerous streams with potential quite close to the town. Many gold- or gem-bearing creeks in these ranges lie close to the New England Highway only a short distance from a town, and these make ideal, and often very rewarding, places for beginners to fossick.

Braidwood district

Braidwood is a fine, historic town, well preserved and retaining all the atmosphere of the hectic gold-rush days. Not far from the town flows the Shoalhaven River, one of the most important rivers of the south coast. In the early days the Shoalhaven was a source of prolific gold finds, and fossickers today still consider it a good prospect, particularly in the upper reaches. Harnessed by a massive water storage system near Nowra, the Shoalhaven is now more important for its dams and lakes, which will one day supply water to most of the south coast towns. But the upper reaches of the river, near Braidwood, are relatively unaffected by such development and good "colour" is often obtained from this big river and its highland tributaries.

The countryside along this part of the Great Dividing Range is magnificent, and Braidwood is set among lush, rolling hills studded with granite tors and scattered boulders. It is a well-organised town, lying astride the main highway between the south coast and Canberra, and is a good centre from which to begin fossicking expeditions in a number of different directions. While there are camping facilities in most of the villages near fossicking sites, few have anything more than a local store, whereas Braidwood, located centrally among the gold areas, is an ideal focal point with every requirement for fossicking families.

The most productive of the sites near Braidwood is at Majors Creek, on the Braidwood to Captains Flat road. Jembaicumbene and Duck creeks and the Shoalhaven River itself are all potential fossicking areas, but literally any creek in the immediate vicinity would be worth trying. The area from Jinglemoney south along the river has many access points and this has been one of the most prolific stretches of the river in the past. Roads are mostly gravel and easily handled by normal family vehicles.

Araluen Creek, a tributary of the Deua or Moruya River, is another stream with good potential. Fossicking has produced good results in these streams, although the river cannot be panned where it runs through the Deua National Park. In the gold-rush days, the Araluen Valley was well known by prospectors and also by bushrangers who were fairly active in the area, obtaining their gold by somewhat easier means than the diggers. In 1986 a large nugget was found near Araluen.

The Mongarlowe River, which eventually runs into the Shoalhaven River near Charleyong, is another of the highland streams that has promise for fossickers. The village of Mongarlowe is less than 20 kilometres to the north-east of Braidwood, and stretches of the river to the north and south of this town are well worth a try, as good gold has been recovered by panning. Many of the streams that empty into the Shoalhaven River in this area could have gold, for a little to the north, near Oallen Ford and Nerriga, gold has been a feature of considerable activity

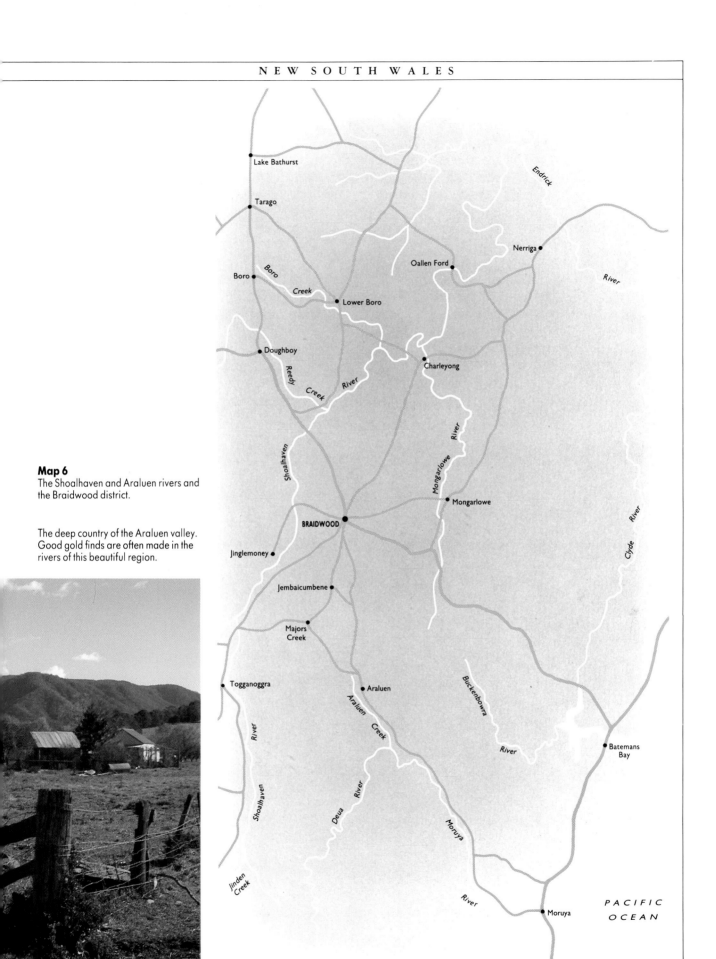

Map 6
The Shoalhaven and Araluen rivers and
the Braidwood district.

The deep country of the Araluen valley.
Good gold finds are often made in the
rivers of this beautiful region.

Lake Bathurst

Tarago

Endrick

Nerriga

Oallen Ford

Boro

Boro

Creek

Lower Boro

River

Doughboy

Charleyong

Reedy

Creek

River

Shoalhaven

Mongarlowe

River

Mongarlowe

BRAIDWOOD

Clyde

River

Jinglemoney

Jembaicumbene

Majors
Creek

Togganoggra

Araluen

Buckenbowra

Araluen

Creek

River

Batemans
Bay

River

Shoalhaven

Deua

River

Moruya

Jinden
Creek

River

Moruya

PACIFIC
OCEAN

0   5   10   15   20 km

both by amateur and professional prospectors.

Since gold can be carried long distances by rivers and streams, particularly by fast-flowing water, any stretch of the Shoalhaven, from its headwaters south of Braidwood to its entry into the spectacular Morton National Park, could produce results. Many of these areas, particularly to the north of Braidwood, are relatively isolated and rugged, so the streams will not have been worked as frequently as those with easier access. However, roads are rough in places and a four-wheel-drive vehicle will be required where access is off the gravel roads.

Tibooburra and Milparinka

In complete contrast to most of the gold-bearing regions closer to the coast, the Tibooburra–Milparinka region is mostly flat, arid country where creeks and streams are dry for most of the year. Indeed, one might be forgiven for questioning the name Milparinka, which is an Aboriginal word meaning "water may be found here"! It refers, of course, to a well-established waterhole, which may well have saved Charles Sturt and his exploration party when they camped near the site of the present town in 1845.

Map 7
Tibooburra and Milparinka.

Small nuggets are still found beneath the dry, dusty soil of the Tibooburra region. Metal detectors are best for this environment.
PHOTO BOB GREIG, MINERS' DEN.

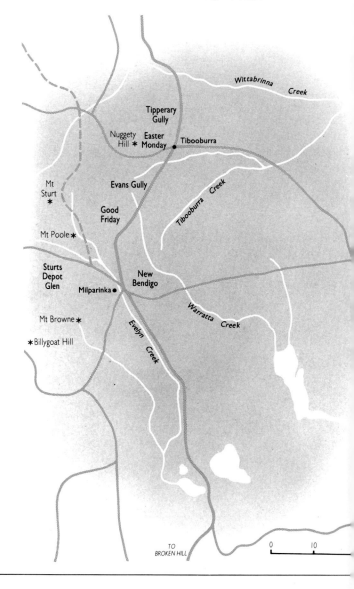

Milparinka is located just off the Silver City Highway, around 290 kilometres north of Broken Hill, while Tibooburra is a further 41 kilometres to the north. Lack of water restricts fossicking in these areas to either dry blowing or the use of metal detectors. Quite good finds have been made with both methods, although for amateurs, the metal detector is easier and more likely to produce results. The highway is accessible to normal family vehicles although it can be rough in patches after heavy rain. The tracks and side roads that lead to fossicking areas mostly demand the use of a four-wheel-drive vehicle.

The main areas where gold has been recovered lie to the west of the highway, although one, known as New Bendigo, lies on the eastern side along Warratta Creek. Sturts Depot Glen and Billygoat Hill, in the Mount Browne Range, are reached from Milparinka across rough roads to the west of the town. Good Friday and Evans Gully are located from a turnoff along the highway between Milparinka and Tibooburra, while the area immediately to the north of Tibooburra has deposits in the areas known as Tunnel Hill and Easter Monday, Nugget Gully and Tipperary Gully.

Tibooburra has a motel, hotel and caravan park as well as all facilities for fossicking families. Milparinka has only its old pub and a few facilities, and since most supplies are expensive because they have to be brought in, it would be wise to stock up in Broken Hill before setting off along the Silver City Highway. Temperatures in these areas can be extreme and inexperienced travellers should obtain advice from the local police before moving far out of town.

Forbes–Parkes district

Although once one of the most prolific gold-producing areas in New South Wales, the Forbes–Parkes goldfield has produced most of its precious metal from mining. The deep leads in the area, which were formed by secondary deposits, are mostly beyond the reach of amateur miners, for they require shafts to be sunk to considerable depths to reach the alluvial beds. Some of the more important deep leads have been mined to depths of 70 metres and more.

Nevertheless, where there is gold of any description there is potential for amateur prospectors and in this case the best results might be obtained by working the old mine tailings. Miners of earlier days having done the hard work of getting the pay-dirt to the surface, there is scope for fossickers to go through the mullock heaps and tailings again and still get results. Metal detectors would be useful here, although since it is highly unlikely the earlier miners will have left nuggets of any size, the detectors will need to be extra sensitive to pick up small grains and particles of gold that may have slipped through previous workings.

As in any area where gold is found, the streams and creeks are also worth trying. Some surface outcrops have been located in the district and primary deposits of this type usually indicate the presence of alluvial gold in adjacent drainage systems. However, this goldfield is not known for its creek gold, as are places such as Stuart Town and Sofala, farther to the east.

An indication of the extent to which the early miners would pursue the gold can be seen by the huge crater at Peak Hill — the remains of one of the biggest mines in the state. The Forbes–Parkes goldfield lies in a belt

about 50 kilometres long by 10 kilometres wide adjacent to the two towns from which it takes its name. Over the years since it was first worked it has produced no less than 18,900 kilograms of gold. A new project involving open-cut mining is being planned in the vicinity of Parkes, which will bring this field alive again, long after it was thought to be exhausted. So, if it can attract the huge investment of commercial operations, there must also be good potential for amateur prospecting.

Access is by the Newell Highway to the townships of Forbes, Parkes or Peak Hill, and there are reasonable gravel feeder roads leading into the gold regions. Once off these roads, however, access to some areas will be by four-wheel-drive vehicle only, so a certain amount of investigation is a wise precaution before setting out. It will also be necessary in many cases to obtain permission from property owners since the old mine workings are mostly on private property.

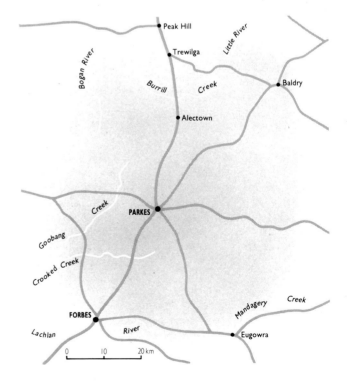

QUEENSLAND

Gold mining is booming in Queensland. Apart from the development of a huge new mine (Kidston) and the discovery of new gold deposits (Pajingo), the current high prices for gold on the world markets and the use of new techniques in mining and ore treatment have caused many of the old fields and mines to be re-examined. In many cases the tailings as well as some alluvial areas are being reworked, and the results are sufficiently encouraging to suggest that Queensland is on the verge of a new gold boom. In the 1985/86 financial year, almost 11,000 kilograms of gold was produced, nearly double that of the previous year.

There are more than 100 goldfields in Queensland, most of which contain alluvial gold, making the northern state a fossicker's paradise. Wherever gold is located, the streams, creeks and rivers in the vicinity are often potential sources of the precious metal. Even fields where most of the gold was recovered from reef deposits have potential for fossickers, both in the tailings of the old mines and in nearby streams, where the erosion of surface outcrops may have formed alluvial deposits.

However, most of the alluvial gold recovered near the surface in past operations has been concentrated in eight specific areas. The Palmer River produced more than half the total amount, Gympie accounted for 16 per cent and Clermont 11 per cent. The other major producing areas were Cape River, Etheridge, Hodgkinson, Rockhampton and Wenlock. Some of these fields are already being reworked by commercial miners, which indicates the potential for amateur fossickers, whose financial outlay need only be minimal.

All but a few of the Queensland goldfields are in the eastern half of the state; most are on or near the spinal ridge of the Great Dividing Range. The exceptions are the deposits near Cloncurry and Croydon which lie more in the central western region. Although the major de-

Left: A huge crater left by early mining activities near Peak Hill. The tailings of old mines in this area are being reworked, and a new mine is about to come on-stream, indicating the potential of the region to produce more gold.

Map 8
The Forbes–Parkes region.

Below: Many of Queensland's mineral fields lie in difficult country. Access often requires four-wheel-drive vehicles.
PHOTO TOM BUDDEN.

posits are in localised areas, the gold belt stretches the length of the state from Warwick, near the New South Wales border, to the tip of Cape York. An excellent map, produced by the Queensland Department of Mines, provides useful information about gold localities. The Department also produces a booklet titled *Gold Fossicking*. These publications, together with others on specific areas such as the Gympie gold fossicking area, can be obtained at the Queensland Department of Mines, Brisbane.

Warwick district

This goldfield lies on the northernmost extension of the New England Fold Belt and is centred roughly between Warwick and Texas. Good alluvial gold has been found in the area and there are a number of old mines which could provide potential in their mullock heaps and tailings. Access is along the Cunningham Highway to Karara and the junction of the road north to Leyburn. The creeks in the area all have fossicking potential, the most important being Canal Creek, Thanes Creek and Sandy Creek.

Once off the main roads, the going can get rough and four-wheel-drive will be necessary to penetrate the bush to the less accessible creeks and waterways. However, reasonable access to the creeks where they run close to, or under, formed roads is possible at many spots, notably with Sandy Creek, to the west of the Leyburn road, and Canal Creek to the east. Fossicking families with conventional two-wheel-drive vehicles will find plenty of spots to try their luck, although care must be taken not to enter private property without first obtaining the consent of the owner.

Map 9
The major gold-bearing areas near Warwick.

Although off-road access can be difficult, there are some very lucrative streams in this district.

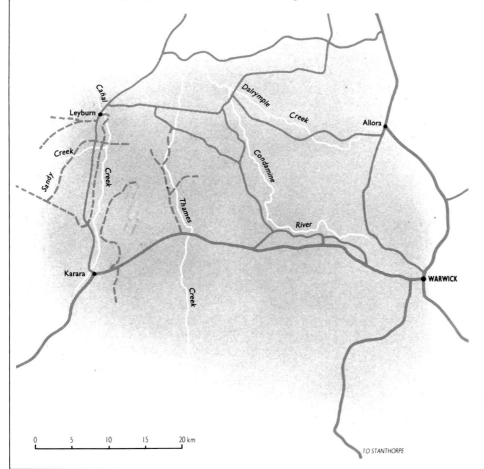

The nearest town to the fossicking areas is Leyburn, but this has only the basic facilities for food and fuel supplies. Warwick, some 80 kilometres away, is the nearest town with accommodation and other requirements. Because this is a relatively isolated area, the potential for making a lucky strike is good. The run out from Warwick is, for the most part, on good roads, so should take no more than an hour. Other deposits have been found in the vicinity of Stanthorpe, around 60 kilometres to the south of Warwick.

Gympie–Rockhampton goldfields

Of all Queensland's many gold strikes, none was more important than that at Gympie. It was not the first discovery of gold, nor was the deposit the biggest, but it came at a critical time in the state's history. Queensland was experiencing a severe financial crisis, and failure of a number of banks, including the new Bank of Queensland, plunged the state into chaos and sent unemployment rocketing. The timely discovery of a big gold deposit at Gympie in 1867 put the Sunshine State back on the rails and ensured continuing prosperity for some years to come. It was the beginning of a golden era, for as prospectors flooded north, more and more fields were discovered and Queensland became the gold centre of the continent, if not the world.

James Nash, a professional prospector from Victoria, first discovered gold near the present site of the Gympie Town Hall in 1867. He had camped in a gully, which now bears his name, on a journey from Nanango to Maryborough. As a matter of routine, he panned the wash-dirt of the gully, and discovered that it was rich with gold. So rich, in fact, that in the ensuing gold rush some 15,000 prospectors worked the area without exhausting the alluvial dirt and recovered gold in startling quantities and sizes, including the "Curtis Nugget" which weighed over 28 kilograms. One of the prime features of the creeks in the Gympie field is the shallow depth at which the gold is found, often in the surface wash-dirt.

An enterprising Queensland Government has set aside one of the major gold-bearing creeks in the Gympie field for the use of amateur fossickers. Since Deep Creek is at the southern entrance to the town, panning for gold can be carried out even by raw beginners without any of the trauma associated with getting into difficult fossicking areas, and the isolation of distant goldfields. Here you can stay in relative luxury in a motel or camping ground and literally walk to the creek. Although results obviously cannot be guaranteed, Deep Creek was one of the most prolific of the Gympie creeks so there is always the chance of finding a missed nugget.

The belt of gold deposits that begins just south of Gympie spreads northwards and westwards along the ranges to cover an extensive area of southern Queensland. To define its limits exactly is difficult since in many areas gold deposits may not yet have been located because of the inaccessible terrain in which they lie. Since this book is concerned with amateur fossicking, which does not generally include mining reefs or deep leads, the only areas mentioned are those which are reasonably accessible and where good alluvial finds have been made.

Nanango is roughly the southernmost tip of this field and Rockhampton the northernmost. Although gold is not commonly found close to the coastline, alluvial gold, including nuggets, has been found near Calliope, not far from the port of Gladstone. The western extremity of the field is at Cracow, although gold recovered from this area has been mostly mined. The concentration of gold within this field varies considerably, but taking into consideration the factors mentioned earlier relating to access and alluvial deposits, the main areas that could be productive for amateur fossickers are Kilkivan, Eidsvold, Degilbo, Bompa, Colo Flat, Palmwoods, Pomona, Calliope, Callide Creek, Mount Cannindah, Barmundoo, Canoona, Cawarral, Moonmeera and Mount Morgan.

The poppet head of an old gold mine stands as a memorial to Gympie's heady goldrush days.

Map 10
The gold bearing region around Gympie.

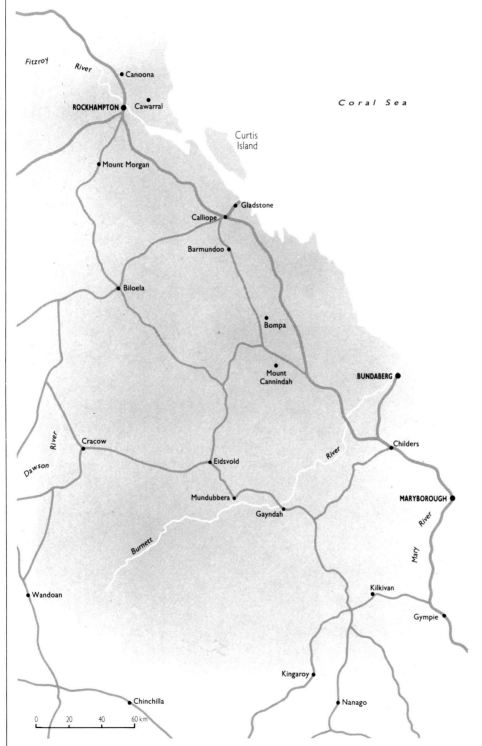

Clermont district

The Clermont district is well known for mining activities of many kinds. Rich copper deposits were worked in the Peak Downs area in the nineteenth century and the Blair Athol open-cut coal mine, 20 kilometres from the town of Clermont, is one of the most important in Queensland, with a seam of coal more than 30 metres thick in places. The area lies in the mineral-rich Bowen Basin, and Blair Athol alone is known to have reserves of over 200 million tonnes of coal located close to the surface.

But it was gold that first brought Clermont to life in 1861, and it is gold that attracts most visitors to the town today. The surrounding district has produced considerable alluvial gold since the early miners moved on and there is doubtless a great deal more waiting to be recovered. To the north of the town, between Clermont and the property of Miclere, are a number of areas where good alluvial gold has been found, while close to the town are Exhibition Creek and McDonalds Flat, from which results have also been obtained.

Apart from the gold in the proximity of the town itself, Clermont lies close to the gemfields around Anakie, well known in Queensland as the gem-fossicker's paradise. An expedition to this area would be well worth while for those interested in any form of fossicking or prospecting, for a major goldfield and a major sapphire field lie within less than 100 kilometres of each other. Full details of fossicking for sapphires in this area are given in Chapter 5.

Map 11
The Clermont district.

Apart from its well-known goldfields, Clermont has one of the most prolific gemstone areas in Queensland at nearby Anakie.
PHOTO VICKI ARMSTRONG.

Charters Towers belt

Probably the largest group of gold deposits in Queensland lie in this region which takes the form of a belt running east–west across the ranges from the coast to the Western Slopes. I have called it the Charters Towers belt since that town lies roughly in the centre, and in any case was the focal point of most goldmining in the region during the gold-rush period. From a fossicker's point of view, it is more of interest for re-working old mine tailings than for panning streams, since much of the gold retrieved from this area was mined from reefs. However, where there are reefs, there is often eluvial gold in the vicinity and sometimes alluvial gold in the streams, so amateurs equipped with metal detectors or panning dishes may easily strike pay-dirt in the surrounding country-side as well as in the mine tips.

Near the coast, gold deposits have been found from east of Mackay to just south of Bowen, mainly in the coastal ranges. Eungella, at the head of one of the most delightful valleys in the state, is well known for its alluvial gold, as is nearby Mount Britton, where good-sized nuggets have been recovered. The district to the north and south of Dittmer, inland from Proserpine, has also been productive in past years. Near the junction of the Sutton, Cape and Burdekin rivers there are a number of areas where good alluvial gold has been recovered, particularly upstream towards the headwaters of the Sellheim River, near Mount McConnell.

In the Ravenswood–Charters Towers areas, a great deal of the gold has also been taken from reefs, although in the northern areas, in the vicinity of the Fanning River and Mingela, good alluvial deposits have been located. Farther west, between Charters Towers and Hughenden, the prolific Cape River cuts under the Flinders Highway near the village of

Right: Most of Charters Towers' fine old buildings are directly related to the boom days, when alluvial gold was found in streams in the centre of the town.

Below right: Tailings, eroded by years of weathering, create moonscapes in the surrounding countryside.

Map 12
The Charters Towers belt.

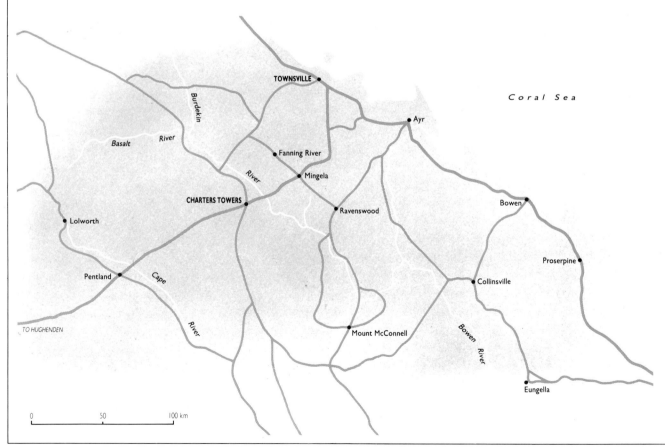

Pentland. Here a turn-off runs north through the Cape River region, winding into the hills near Lolworth and providing access to excellent fossicking grounds. Commercial operations have recommenced on the Cape River, indicating the potential of the area. The road in from the highway is passable, but access to the river may require a four-wheel-drive as well as permission from the owners of riverfront land.

Cloncurry district

Most of the gold deposits located near Cloncurry have been to the south of the town, the alluvial finds being to the east or south-east of Malbon, on the Dajarra road, in country which gives rise to the headwaters of the Cloncurry River. Other alluvial deposits have been found to the north of nearby Mount Isa and near Kajabbi, to the north of Cloncurry, in the headwaters of the Leichhardt River. These are mostly isolated areas and it is hard to imagine amateur fossickers braving the arid, rough conditions in this part of the country.

Georgetown–Croydon district

Although alluvial gold was found in Croydon in 1885, most of the gold recovered from the surrounding area was mined from reefs and leads. It was a very prolific field and during its heyday, between 1887 and 1906, was one of the greatest gold producers in Queensland, being topped only by Charters Towers, Gympie and Mount Morgan. Commercial operations virtually ended in 1923 but there is renewed interest in the area.

Map 13
Croydon, Georgetown and its gold-bearing district.

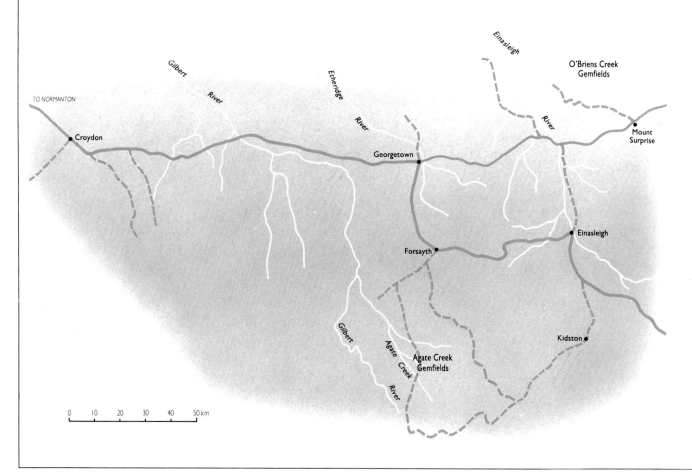

The big rivers of the north and their myriad tributaries and creeks provide good fossicking for those who have the means to gain access.

The best fossicking here is in the tailings of the old mine workings — which are scattered across the countryside, mainly to the south and east of the town — in an area known as The Springs. Metal detectors are invaluable in places such as this where water is scarce. Roads are rough and a four-wheel-drive vehicle will be necessary to reach the main gold-fields. The climate is harsh at certain times of the year so inexperienced fossickers should not work this area unless accompanied by someone who knows the country.

Georgetown lies 53 kilometres eastward along the same highway. Here there is a broader mix of primary and secondary deposits. The whole area south of the town, which is reached by a dirt road that curves through Forsayth and Einasleigh, is dotted with old mine sites and potential fossicking areas. The Einasleigh, Etheridge and Gilbert rivers have countless creeks running into them in this area, any one of which could be carrying gold. However, getting off the main road and into the abandoned mines or the creeks wil require a four-wheel-drive vehicle.

From Forsayth or Einasleigh, minor roads and dirt tracks lead to distant mining areas such as Gilberton and nearby Agate Creek gemfields. This is a fossickers' paradise with a choice of gold or gemstones scattered across the countryside in an area which is so remote it could never be thoroughly worked out. Neither of these villages has any but the basic necessities and Croydon has little more. But Georgetown has two caravan parks and a motel/resort, as well as a number of shops and facilities.

Palmer River and the northern goldfields

The northern goldfields are scattered over such a wide area that it would be virtually impossible to cover them all in a book such as this. Good finds have been made in the coastal mountains between Innisfail and the Atherton Tablelands, just as there have been profitable strikes north of Coen, on the Cape York Peninsula. Inland from Cooktown lies the most famous of all Queensland fields, which incorporates a number of gold-producing areas known collectively as the Palmer River goldfield.

Near the coast, good access is available to fields at the foot of Queensland's highest mountain, Mount Bartle Frere. From the Bruce Highway, just north of Innisfail, a 4 kilometre road leads to the settlement of Bartle Frere. This provides moderate access to the creeks that feed the headwaters of the Johnstone and Russell rivers, and in which the first alluvial gold in the area was found. Historic names such as Christmas Creek, Sandy Creek and Jordan Creek, Tinaroo and Mount Peter are scattered around these ranges which in their time have witnessed the frantic activities of several rushes for buried treasures. Now much of the Bellenden Ker Range is a national park and fossicking is not permitted.

Map 14
The northern goldfields.

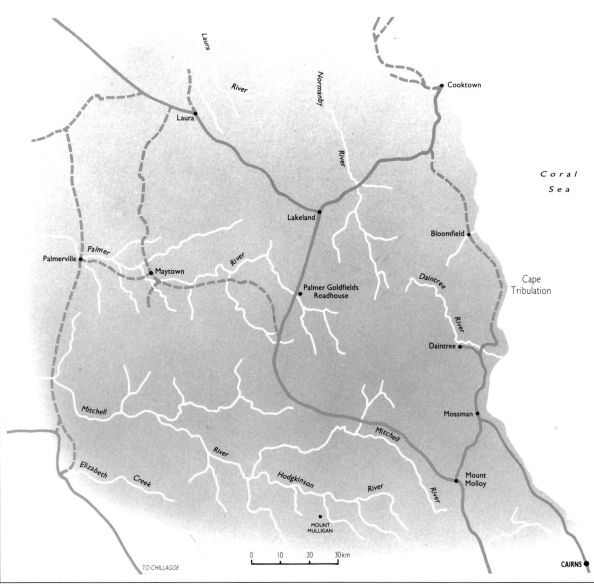

An amateur prospector set up to work a creek in the deep forest of the Cape York Peninsula country.

PHOTO TOM BUDDEN.

Inland, the fields stretch northwards from Chillagoe and Mareeba through the wild, rugged region that was once known as Mulligan country after the Irish digger who worked the area then made northwards to find the huge field along the Palmer River. From Dimbulah, on the Chillagoe road, access is possible along a rough road to Mount Mulligan, in the heart of this impenetrable country. Looking north over the endless timbered ranges it seems inconceivable that men could have traversed such inhospitable terrain through extremes of climate, without any mechanical assistance, often without food or facilities, and with the constant threat of attack from angry Aborigines. From the rich goldfield of the Hodgkinson River they walked north, suffering extreme privations and covering endless kilometres up and down steep ridges and through deep valleys to reach their destination on the Palmer River. Such was the lure of gold.

Access to the southern areas is still almost impossible. Even with four-wheel-drive vehicles, amateur fossickers would be well advised to leave this wilderness country alone and press on to the northern region where access into the Palmer River region, although difficult, is possible with a four-wheel-drive vehicle. The Peninsula Developmental Road branches northwestwards at Mount Molloy, itself an old mining village, and heads towards Lakeland, Laura, Coen and eventually Cape York. There are numerous old gold mines and sites in this area, but they are virtually impenetrable, even with a suitable vehicle.

The principal fossicking area with access, albeit fairly rough and requiring a four-wheel-drive vehicle for safety, is the famous Palmer River goldfield. Located about one-third of the way across the Cape York Peninsula, inland from Bloomfield and Cooktown, this region was the scene of a frenzied gold rush in the late nineteenth century. Literally thousands of Europeans and Chinese diggers fought their way through the dense bush to wash and hack a fortune in gold from the Palmer River and its surrounding area. The tales of this gold rush are an integral part of Australia's history, for in no other goldfield in the nation was the gold won against greater odds and in more horrendous conditions.

James Venture Mulligan, a prospector whose name is perpetuated in nearby Mount Mulligan, set off, in June 1873, with a party of six others from the goldfields of Georgetown to search for gold in the northern wilderness country. After a tortuous journey through the virtually unexplored country they reached the Palmer River and set up camp. Within three weeks the party had recovered over 3 kilograms of gold. The news of Mulligan's strike spread across the country, triggering off a huge rush. By September 1873 the first of the thousands had set up camp on the present site of Palmerville, and soon afterwards more than 500 miners were working the lucrative river. By mid-November a makeshift port had been established at Cooktown and a track blazed across the rugged counryside to service the new goldfield. The frantic rush grew to the point where, by 1877, almost 20,000 diggers were working the area. Of these, about 17,000 were Chinese, and only 1400 Europeans.

The intensity of the activity quickly depleted the alluvial gold in the streams and the miners moved on as new discoveries were made. Following the abandonment of Palmerville, Maytown became the centre of activity. Then gold was discovered in the Hodgkinson River, farther to the south, and many of the European miners moved camp, leaving much of the Palmer field to the Chinese. Conditions in all centres were nothing short of horrific, with disease, tropical monsoons and hostile Aborigines taking their toll. Despite this, the goldfields of the ''Peninsula'' were among the most lucrative in Queensland and played an important part in the state's prosperity.

Limestone outcrops beside the old Laura-to-Palmerville coach road. Access to the Palmer River goldfields is through rough but spectacular country.
PHOTO TOM BUDDEN.

Remains of a dream. Rusting boilers and machinery mark the site of one of Australia's most historic goldfields at Maytown on the Palmer River. Good results are still obtained from the creeks in this district.
PHOTO TOM BUDDEN.

When the alluvial gold ran out, reef gold was mined, but there was insufficient to keep up the hectic pace and as the 1880s came to an end the fortunes of the Palmer goldfields began a long decline. The towns created by the boom became ghost towns. The bush crept back to reclaim the areas that had been cleared and covered the scars left by the miners. The roads were absorbed into the undergrowth and the creeks and rivers ran freely again, unhindered by dams and sluicing pools. Little was left as a legacy of one of the major events in Australia's early history save a few rusting machines and slowly disintegrating ruins.

Although the decline of the Palmer goldfield began almost a century ago, gold is still being taken from this quite remarkable source. One mine, complete with stamper, worked the reef deposits until 1976, and amateur fossickers have always reaped the rewards of making the hard trek in from the highway to pan the local streams. In recent years commercial mining companies have shown a renewed interest in the area and another gold boom — albeit only a 'mini' boom — seems to be under way. Amateur prospectors and fossickers are in the vanguard of the rush to re-examine and rework much of the leftovers from the initial gold rush.

Access into the Palmer River area is still limited, and rough roads, creek crossings and rocky river beds require the use of a four-wheel-drive vehicle. The most widely used route involves a journey of about 280 kilometres from Cairns, firstly along the Peninsula Developmental Road and then through the bush along about 80 kilometres of rough track known locally as Mulligans Highway. The turn-off for this track is about 5 kilometres north of Reedy St George crossing and leads to Maytown, Granite Creek, Cannibal Creek and Uhrstown. The track also continues on to Palmerville along the old Maytown to Laura coach road, but this is very rough indeed, and not recommended except for experienced bush drivers.

A visit to Maytown is interesting from a historical point of view, for the remains of much of the old gold town are still to be seen. Restoration and preservation work is in progress in parts of the district, and the Palmer River Historic Preservation Society has erected a replica of a miner's hut which includes a display of relics found in the area. Work is in progress restoring one of the machinery complexes — the Comet machine and boiler installation. An excellent map is produced by the Queensland Department of Mines showing the layout of the Maytown area together with details of the buildings, streets and principal mines, and a brief history of the region.

There is still undiscovered gold in the streams of the surrounding countryside, if not in the Palmer River itself. Even the diligent Chinese could not prospect this area completely and good finds are frequently made in creeks and streams off the beaten track. There are no facilities whatsoever in the whole goldfield area, so the only way to live is under basic bush camping conditions, which means that all supplies must be brought in. The nearest place for any requirements is the Palmer River Roadhouse, on the main north road.

There are other tracks into the wilderness areas of both the Palmer River goldfields and other goldfields along the Cape York Peninsula, but these are often extremely rough and lead into seemingly impenetrable bush. Amateur fossickers would, therefore, be ill advised to attempt to enter them.

VICTORIA

No state in Australia has more gold deposits than Victoria. With the exception of the far west of the state, there are goldfields scattered across almost every region, from the Grampian Mountains to the New South Wales border. Gold has been recovered in huge quantities from a wide diversity of geographical locations, the main concentrations being in the central western region around Bendigo, Ballarat and Maryborough. Much of the gold has been recovered from deep leads in old river beds, but a great deal has also been won the hard way — from the quartz veins of gold-bearing reefs. Between 1851 and 1900, Victoria was the largest producer of gold in Australia and the industry played a major part in the development of the state. Since the first gold was discovered in 1851, an estimated $10,000 million of the precious metal has been recovered.

Although like most other states, other than Western Australia, the boom did not last and the goldfields were gradually abandoned, Victoria has maintained a few viable commercial mining or dredging operations to this day. Here, as in other states, the escalating price of gold and the advent of metal detectors has started a minor boom as thousands of amateurs rework the old mine tailings and dumps.

A unique feature of the Victorian goldfields is the high proportion of gold found in the form of nuggets. Although other states have produced nuggets, some of them larger than those found in Victoria, nowhere else has there been such a proliferation of nuggets. The district around Dunolly has produced some 126 nuggets, and nearby Rheola 98. The central-west region would appear to be the most likely place to encounter this form of gold, for although far behind Dunolly and Rheola, other nugget finds in the area include Wedderburn 40, Ballarat 38 and Bendigo 23.

An interesting feature of these finds is that more than half the nuggets exceeded 2.8 kilograms (100 ounces) in weight. Many were flat on one side and rounded on the other, due to the abrasive action of material passing over the nugget, rather than rolling it along. The use of metal detectors has resulted in many good nugget finds, since this is the type of fossicking for which the electronic detector is particularly suited.

Auriferous gravels occur on the surface at virtually every goldfield in Victoria, and although all have been well worked, both by the original diggers and by later fossickers, good recoveries are constantly being reported by amateur fossickers. Most of the gold is found in old workings, and since a large percentage of the state's ground surface is pockmarked with old diggings, there is no shortage of fossicking areas.

Since the goldfields extend over vast areas of central and eastern Victoria, it is not possible to list all potential fossicking spots. As with other states, the details provided in this section deal with the goldfields most likely to produce results for family fossickers, taking into consideration the potential of the field, the access for normal family vehicles and the ease of obtaining gold. Where recovery of gold requires shaft mining or the use of heavy equipment to break up quartz or rock, it is better left to experts.

Most of the goldfields in Victoria lie in belts running roughly north to south. To make identification easier, the locations given here are described in belts rather than in local districts.

One of the biggest gold strikes in Victoria was made here, on the gentle slopes of Mount Alexander.

Victoria's goldrush past is kept alive in
the exciting recreation of a gold town at
Sovereign Hill.

The Pyrenees belt

Named after their magnificent counterparts in Spain, the Pyrenees are a delight to visit and fossick in, whether or not the results are profitable. The gold-bearing region is centred on the town of St Arnaud, 265 kilometres north-west of Melbourne, and stretches across the adjacent towns of Avoca and Beaufort to the south. Access is therefore very easy along the Sunraysia Highway from Ballarat, and most of the feeder roads have good surfaces.

However, this is a lush rural area and access to the goldfields, the streams and creeks depends on obtaining permission from the owners of the land. Panning the creeks with headwaters in the Pyrenees can be very rewarding, for much of the gold recovered from this area was alluvial, some from deep leads but much from surface gravels.

Most of the workings are to the west of the Avoca River and near locations known as Moonambel, Redbank, Lambing Flat, Stuart Mill, Argyle Gully, Charcoal Gully, Mogs Rush, Lamplough, Amphitheatre and Beaufort.

Map 15
The Pyrenees district of Central Victoria.

Although not actually in the Pyrenees belt, adjacent Stawell was the centre of big gold mining operations. This open cut mine is now one of the town's tourist attractions.

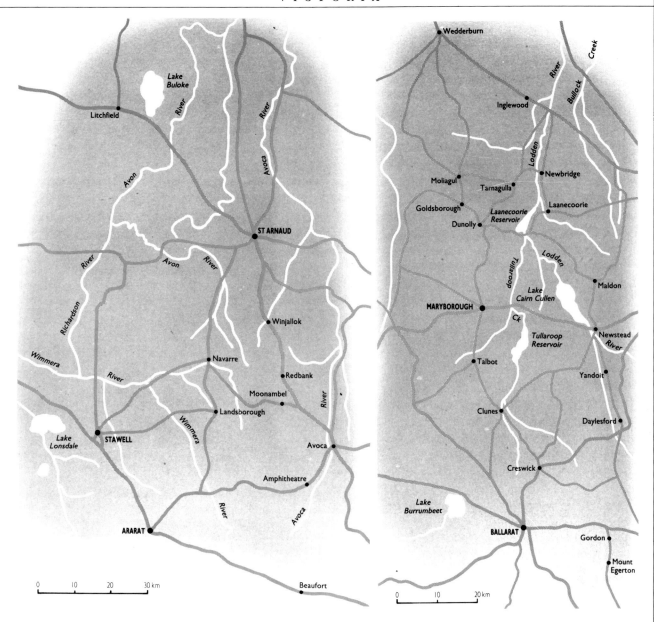

Map 16
The Ballarat–Wedderburn belt.

Ballarat–Wedderburn belt

This is the area best known for its nuggets, for it is the centre of a huge goldfield which has produced many of the biggest nuggets recovered in Australia, among them the Welcome nugget which weighed about 62.8 kilograms. Access to most of the goldfields in this district is easy, indeed some are clearly marked as sightseeing attractions for tourists. Apart from the fields at Ballarat itself, the more prominent fields in the district are Maryborough, 66 kilometres north of Ballarat; Clunes, 27 kilometres north of Ballarat; Creswick, 19.2 kilometres north of Ballarat; Dunolly, 19 kilometres north of Maryborough; Gordon, 20 kilometres east of Ballarat; and Wedderburn, 72 kilometres north of Maryborough.

Other areas where gold has been worked are Rokewood, Inglewood, Berringa, Scarsdale, Rheola, Piggoreet, Pitfield Plains, Nuggetty Gully, Creswick Creek, Talbot, Moliagul and Tarnagulla. Moliagul and Poseidon, close to Tarnagulla, are renowned for a proliferation of nuggets, many of them just below the surface.

Bendigo–Blackwood belt

One of the longest surviving big goldfields, the Bendigo fields were still producing gold less than 30 years ago, having been in the forefront of the Victorian gold rushes of the 1850s. There are few goldfields in the world that consistently produced gold for over a century, and even fewer in Australia, so it is small wonder that Bendigo became the focal point of goldmining history in Victoria. In its time one of the biggest of the central-western goldfields, it is still the centre of a great deal of activity, nowadays mostly from amateur fossickers.

As with all the Victorian goldfields, the Bendigo field covers a vast area, most of which is easily accessible from the city itself. Apart from the diggings in the immediate vicinity of Bendigo, the major fields in this belt are Castlemaine, Trentham, Blackwood, Chewton, Fryerstown, Spring Gully, Ellesmere (Fosterville), Spicer Gully, Little Hampton and Blakeville.

Steiglitz–Maldon belt

The most southerly of the western goldfields, the Steiglitz belt runs north through Daylesford to Maldon in the spa region of the state, where mineral waters were frequently encountered by miners. The main gold-

Map 17
Major gold-bearing fields in the Bendigo–Heathcote region.

Old workings pockmark the Bendigo countryside. Mullock heaps like this provide good fossicking, especially for metal detectors.

Above right: The Central Deborah Mine produced gold for 103 years and is now a living museum.

bearing centres in this belt are Steiglitz, Egerton, Gordon, Daylesford, Yandoit, Maldon, Elaine, Newstead, Blakeville, Sutherlands Creek and Newchum Gully.

Heathcote–Rushworth belt

Almost due north of Melbourne, and only about 100 kilometres along the Northern Highway, Heathcote is one of the first gold towns encountered by visitors heading towards the "Golden Centre" at Bendigo. Mullock heaps, weathered and eroded by the elements, create weird patterns beside the road, relics of the upheaval that tore the countryside apart in the mid-nineteenth century. The belt of gold-bearing terrain between Heathcote and Rushworth is a good fossicking area since much of the gold retrieved from this area has been close to the surface. McIvor Creek and the gullies to the westward have produced considerable gold including some sizeable nuggets.

Main workings on this belt, apart from those in the immediate vicinity of Heathcote, were at McIvor Creek, Costerfield, Redcastle, Whroo, Balaclava Hill and Rushworth.

Mount Baw Baw belt

From its beginnings high on the Cape York Peninsula to its termination near the Grampians, the Great Dividing Range is a paradise for gold hunters. There are few stretches that will not yield gold in some form, either in the streams and creeks that run through deep valleys, or in the outcropping rocks with their weathered quartz veins. The Victorian section is no exception, for from the moment the spinal ridge crosses the New South Wales border, it produces gold, first in the highlands, and then in the rivers that carry off the mountain sediment.

In the vicinity of Mount Baw Baw, the most southerly point of the range, some of the most productive gold country lies waiting for fossickers. It proved its worth with the prospectors of the gold-rush era and on many occasions since has proved that there is still plenty left for latecomers. It is ideal country for family fossickers since the region is one of great beauty; just being there is a reward in itself. Finding gold is the icing on the cake.

Gold has been discovered and mined in many locations in these mountains, so most have passable access, although like all mountainous regions, common sense and care are necessary in some of the more isolated and rugged areas. Probably the best access from Melbourne is through Warburton and Noojee, although most of the roads leading north from the Latrobe Valley provide a means of entering the area.

Broadly speaking, this belt extends from around Neerim, north across the top of the ranges almost to Mansfield, and east past Walhalla. Innumerable finds have been made in this area, some well known, such as the huge deposits at Walhalla. But it would be safe to say that almost any mountain stream in this region would have potential for alluvial gold, the Tanjil River having already proved a rich source.

The major areas where gold has been located in this region are Crossover, Neerim, Tanjil, Upper Tarago River, Red Hill Creek, Crossover Creek, Shady Creek, Frenchmans Gully, Jacksons Gully, Dead Horse Creek, Live Horse Creek, Tanjil River, Russells Creek, Hill End, Pheasant Creek, Walhalla, Woods Point, Gaffneys Creek, Loch Fyne, Matlock, Toombom, Enochs Point, Donellys Creek, Fultons Creek, Alexandra and Yea.

Eastern Highlands belt

This is one of the most extensive belts in Victoria, although the deposits are fairly widely scattered. It extends from Gippsland, in the foothills behind Bairnsdale, and sweeps across the Great Dividing Range in a swathe about 80 kilometres wide to the rich deposits near Beechworth and Chiltern. It takes in the Omeo region of the Snowy Mountains, the Dargo and Bogong high plains, the peaks around Mount Featherstone and the lush slopes of the Ovens Valley. In winter much of the area is snow-covered; in summer it is a carpet of wildflowers. Once again, the aesthetic rewards are always rich, even if the gold jar remains empty.

But it should not remain empty, for like the central-western regions, this was one of the richest of all Victorian gold belts and because of the more difficult access and the greater effort required to get to the locations, the area has not been worked as much as the more accessible lowlands. Of all the Victorian goldfields the highlands provide the most exciting and most promising adventure for amateur fossickers.

The settlement at Walhalla is synonymous with gold. It lies in a valley deep in the Mount Baw Baw region.

Map 18
The Mount Baw Baw region.

The striking ranges in the vicinity of Omeo, once the centre of a major gold rush.

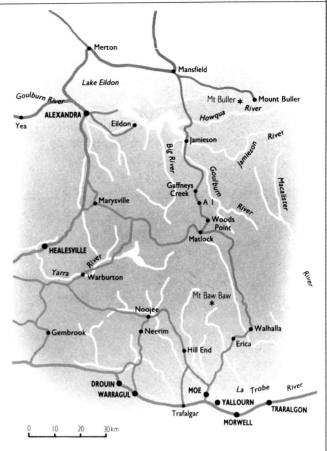

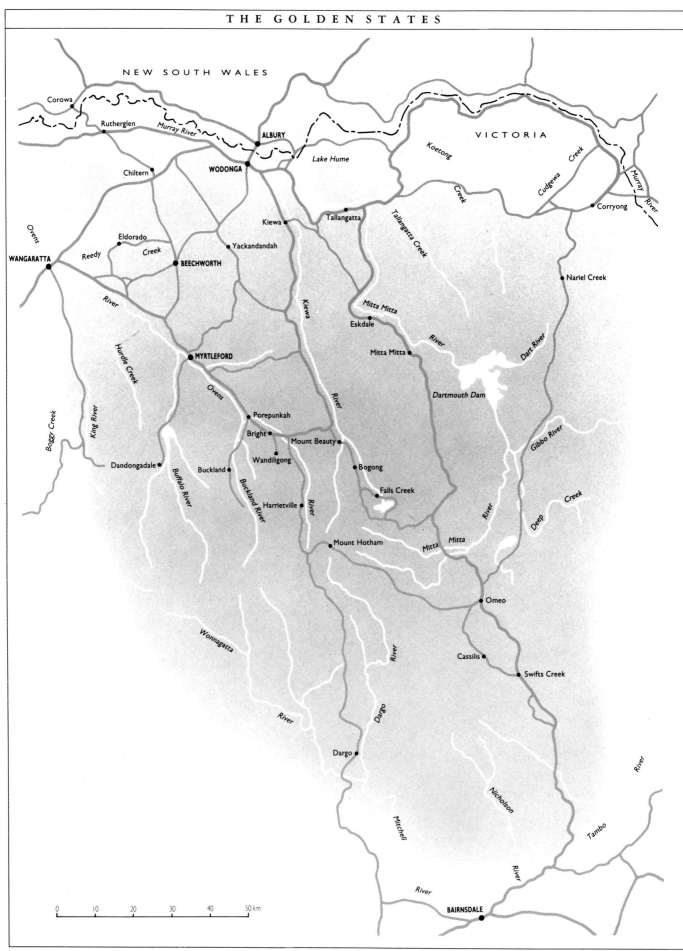

NEW SOUTH WALES

Corowa

Rutherglen

Murray River

VICTORIA

ALBURY

Chiltern

WODONGA

Lake Hume

Koetong

Creek

Cudgewa Creek

Murray River

Corryong

Kiewa

Tallangatta

Ovens

Reedy

Eldorado

Creek

Yackandandah

Tallangatta Creek

Nariel Creek

WANGARATTA

BEECHWORTH

River

Kiewa

Mitta Mitta

Eskdale

Hurdle Creek

MYRTLEFORD

Mitta Mitta

River

Dart River

King River

Ovens

Boggy Creek

Porepunkah

Bright

Mount Beauty

River

Dartmouth Dam

Wandiligong

Dandongadale

Buckland

Buffalo River

Buckland River

Harrietville

River

Bogong

Falls Creek

Gibbo River

Deep

Creek

Mount Hotham

Mitta Mitta

River

Wonnagatta

River

Dargo

River

Omeo

Cassilis

Swifts Creek

Dargo

Nicholson

Mitchell

River

Tambo

River

River

0 10 20 30 40 50 km

BAIRNSDALE

Map 19
The Eastern Highlands belt.

Beechworth's origins go back to the heady days of the early gold rushes. Many of its fine old buildings are preserved for their historic value, and more than thirty are classified by the National Trust.

There are far too many locations where gold can be found to be able to list them all here. But the following areas have all at some time produced significant returns and will provide good fossicking grounds for families looking for fun fossicking: Ovens River, Morses Creek, Buckland Valley, Myrtleford, Murmungee, Harrietville, Wandiligong, Freeburgh, Bright, Upper Dargo, Dargo River, Beechworth, Porepunkah, Eldorado, Hurdle Flat, Hillsborough, Twists Creek (Yackandandah), Chiltern, Rutherglen, Bethanga, Omeo, Mitta Mitta River, Saltpetre Creek, Cassilis, Tallandoon, Glen Wills, Little Snowy Creek, Mitta Mitta, Granite Flat, Lightning Creek, Eskdale and Gibbo River.

CHAPTER FOUR

GOLD
IN OTHER STATES

SOUTH AUSTRALIA

The goldfields of South Australia are much smaller and of less significance than those of the eastern states. The history of goldmining in this state has been one of spasmodic strikes and rushes, with promising results for a while, then sudden depletion of the metal and closing of the mines. The first recorded production from the Victoria Mine near Castambul epitomises this trend. Rich specimens discovered in 1846 caused enormous excitement both among prospectors and on the stock exchange, where the mining company's shares jumped from £2 to £30. But the strike failed to live up to expectations and the mine produced less than a kilogram of bullion.

When the prolific goldfields of Victoria and New South Wales were discovered, a mass migration from South Australia began. Concerned that the state's manpower would be drained to a dangerous level, the South Australian Government offered a £1,000 reward for the discovery of a payable goldfield in the state. The following year this reward was claimed by a man named Chapman who located good deposits of gold near Echunga, only 16 kilometres from the city of Adelaide and now a few

PHOTO SCOTT CAMERON, COURTESY NSW DEPARTMENT OF MINERAL RESOURCES.

Opposite: Access to South Australia's goldfields can be difficult, particularly in the northern regions where only four-wheel-drive vehicles can cope with the rough roads. *PHOTO SHIRLEY HANKE.*

kilometres beyond its present south-eastern suburbs.

Immediately, a gold rush began and the area was overrun by prospectors frenziedly sluicing the streams and burrowing into the countryside. More and more deposits came to light in neighbouring regions, and the Echunga goldfields rapidly developed into a very productive hive of activity. Now-famous names such as Donkey Gully, Chapmans Gully and Jupiter Creek were on everyone's lips as the new field exceeded all expectation. By 1871 it had produced 2600 kilograms of gold which, although not big in comparison with some of the huge strikes in the eastern states, was a considerable success for the relatively small field in the Adelaide hills.

More and more discoveries were made as gold fever swept through South Australia. A prospecting party discovered alluvial gold in Spike Gully, south of Sandy Creek, and the Barossa goldfield was born. Within three years this field was to produce almost 1500 kilograms of gold. Similar strikes were made at Birdwood, Ulooloo and Woodside in 1870 and 1871. It seemed that South Australia, the outsider in the national gold stakes, was pulling back the rest of the field.

Then came a lull, with only sporadic and relatively insignificant discoveries, until the late 1880s when numerous strikes were made in the hills to the north-east of Peterborough. The completion of the railway from Peterborough to Cockburn, on the New South Wales border, provided access to previously desolate areas and prospectors moved in quickly, making good strikes at Waukaringa, Manna Hill, Teetulpa and Wadnaminga. Once again, however, the exhilaration was short-lived and the gold quickly exhausted. The Teetulpa deposit, which was worked by more than 5000 miners, ran out within three years, after producing some 2700 kilograms of gold.

The main goldfield regions in South Australia fall into three principal zones — the group close to Adelaide, the north-eastern fields, and a handful of other sites widely scattered across the central region in a band encompassing Lake Gairdner, Lake Torrens and Lake Frome. Since the latter zone covers mostly arid, inhospitable country not suited to amateur fossicking, the areas described will be confined mainly to the first two zones.

Map 20
Major goldfields close to Adelaide.

Adelaide (east)

It is difficult to set precise limits for this area, since the widely scattered sites where gold has been discovered tend to merge with one another. But by taking the area immediately to the east of Adelaide, between Balhannah in the south and Mount Pleasant in the north, most of the large groupings of goldmining areas are comfortably encompassed. Again, there are many scattered areas within this zone that have produced gold in the past, but since it is impossible to cover every creek, only the major fields are mentioned. The South Australian Department of Mines and Energy produces excellent maps covering most gold-bearing areas and these should be obtained by families setting out on a serious fossicking expedition. Books detailing the local fossicking areas are available from the ''Miners' Den'' shop at Stepney.

Probably the best route into this area is via the River Torrens Gorge Road direct to Birdwood, from where access is possible to goldfields both to the north and south. From Birdwood, a 3 kilometre run along the Cromer Road leads to the first of the mined areas — Hynes Diggings, located on the right-hand side of the road. Adjoining these workings are the Blumberg Diggings and numerous other old workings beside and behind Hynes. About 2 kilometres further along the Cromer Road, on the left-hand side, are the extensive workings of Bennetts Diggings.

Mine shafts are dotted all over the countryside together with their piles of tailings or mullock heaps. To the east, accessible along Lucky Hit Road, are more workings. Cromer Conservation Park and Forest Reserves are located to the north of most of the diggings, and some old mines are located inside the park itself. Fossicking may not be carried out in the park; in forest reserves there are signs to indicate where fossicking is permitted. Most of the workings are on private land and permission must be obtained to enter and work these areas. Access roads are generally passable and although Birdwood has only limited facilities, the city of Adelaide is less than 30 kilometres away so accommodation and other requirements are always close at hand.

Streams of the Fleurieu Peninsula were once the source of considerable alluvial gold. Nowadays they help produce a different form of treasure — the golden grape of South Australia's extensive wine industry.

To the south of Birdwood, on the Mount Torrens road, there are scattered diggings in a number of areas, the most prominent being the Mount Torrens field. Access is relatively easy along side roads which run off the main road at a number of spots. However, this is once again mostly private land and permission must be obtained to work the area, which is ideally suited to the use of metal detectors.

From Mount Torrens, a road leads directly to a number of goldfields which have in the past produced prolific returns. Lobethal is the first of these, only a 7 kilometre drive from Mount Torrens, and Forest Glen goldfields is about the same distance beyond Lobethal. The famous Bird-in-Hand mine is located to the south-east, and the best access here is through Woodside. All these fields are within a few kilometres of each other and all are predominantly reef or lead fields, making the use of metal detectors the best fossicking technique.

The River Torrens, which winds its way tortuously through the hills to the east of Adelaide, is the location for most of the alluvial deposits in this area. Almost any of the dozens of creeks which feed the Torrens provide potential for panning, most of the previous finds coming from the creeks on the south side of the river, upstream from Kangaroo Creek reservoir.

Adelaide (north-east)

Once again a good starting place for this region is Birdwood, for the area to the north of this town is a mass of goldfields. Beginning with the Hynes Diggings, mentioned earlier, and continuing north, there are old mines scattered across the countryside in literally dozens of different locations. Watts Gully diggings produced mostly reef gold, but some alluvial was recovered from nearby areas as well as from the upper reaches of the Torrens River.

Most alluvial gold in this area has been taken from the tributaries and creeks of the South Para River, which, together with its counterpart the North Para River, is a primary tributary of the Gawler River. The road from Gumeracha to Williamstown provides access into the upper reaches of the South Para River and to such well-known goldfields as those around the Bismark and Deloraine mines. Alluvial gold has been found in quantity in the creeks and valleys upstream of the Warren and South Para reservoirs, and also in the famous Barossa goldfield, downstream from the reservoirs. The major activity in this field has been along the Middleton Road, about 4 kilometres south of Sandy Creek, where numerous deep leads and reefs as well as the creeks, have been well worked. Many areas are now restricted by National Park boundaries.

Scattered fields exist farther to the north-east near Tanunda, Angaston and Greenock. These have produced both reef and alluvial gold in the past and are accessible from turn-offs on the Sturt Highway, although most are on private property.

Adelaide (south-east)

This is the area to the south of the South-eastern Highway concentrated mostly around the scene of the first major gold strike in South Australia — Echunga. The extensive deposits of gold run in a south-westerly direction from Hahndorf, across the magnificent rolling South Mount Lofty Ranges almost to McLaren Vale. Within this area are concentrated such

There are remains of old mining settlements in many parts of South Australia. These sites are popular with fossickers using metal detectors, who often find gold missed by the early miners.
PHOTO SHIRLEY HANKE.

The end of a dream. A derelict cottage and its abandoned mine bear silent witness to a digger's struggle to wrest riches from beneath the South Australian soil.
PHOTO SHIRLEY HANKE.

famous mining areas as Donkey Hill, Jupiter Creek, Chapel Hill and Glen Taggart. At the turn of the century these diggings were the major producer of South Australian gold, most of which was alluvial.

Today there are few visible remains of any of the mines, although many of the areas are still favourite spots for fossickers hoping either to find gold that has been missed by the early miners, or a deposit that has remained concealed through the years of frenetic activity. Some parts of the area, in particular the Jupiter Creek diggings, have been placed on the register of State Heritage Items. They are now located in a forest reserve but limited fossicking is permitted.

The best access is by the Strathalbyn Road, which runs through Mylor and Echunga, and on to Macclesfield. Just to the south of Mylor is a turn-off to the left leading to Warrakill Hill. The top end of the extensive goldfield lies between this point and Hahndorf, about 4 or 5 kilometres to the north-east, and there are numerous sites scattered across this area, most of which are located on private property. On the opposite side of Strathalbyn Road are the Chapel Hill diggings, followed by the Jupiter Creek and Glen Taggart goldfields, just south of the Mount Bold reservoir. On the Onkaparinga River above this reservoir good gold recoveries have been made, notably at Seamans Point and Black Rock.

Access to the Jupiter Creek diggings has been made easy for tourists and areas have been set aside for fossicking, subject to conditions displayed on a sign at the entrance to the site. There is a car park on the site of the old township, and a historic trail around the diggings. Access to the area is along Rubbish Dump Road, which can be reached from a number of turn-offs north of Echunga on Strathalbyn Road.

The gold deposits tend to become more scattered beyond the Jupiter Creek field, although three locations to the south have produced good quantities of gold and could provide good fossicking. All have access from Meadows Road, south-west of Fingerboard Corner, and all are associated with Meadows Creek. The first is Blackwood Gully goldfield, which is reached along Peter Creek Road after turning right at Kuitpo Sawmill. The other two, Mount Monster and Willunga goldfields, can be reached by a number of roads, the best probably being the left turn from Meadows Road where it crosses Meadows Creek at Ponders End.

The northern goldfields

From Burra, the remarkable North Mount Lofty Ranges broaden and split into two arms. The main body of the ranges continues its run due north to become the Flinders Ranges, while a less significant arm sweeps around to the north-east past Peterborough and on towards the New South Wales border and Broken Hill. The railway from Peterborough follows this latter arm. When it was opened, in 1887, it provided access to the distant ranges for prospectors, who immediately found gold in a number of areas close to the line.

Manna Hill, Mount Grainger, Teetulpa, Wadnaminga and Waukaringa were the main strikes. Each became prominent goldfields, Teetulpa alone producing about 2500 kilograms of alluvial gold in its first three years of operation. The fields are now deserted except for amateur fossickers and the odd prospector who hacks a living from the dry soil. But all lie relatively close to the Barrier Highway and are within an easy drive of Adelaide, so they provide excellent fossicking in interesting country for amateurs wishing to go a little farther afield. Metal detectors are the best fossicking equipment in most of these areas.

The first field is at Ulooloo, reached along a rough road 9 kilometres north of Hallett, about 30 kilometres north of Burra. The deposits are centred around Coglins Creek, Noltenius Creek and Rubbers Camp Creek, although fossicking could produce results in any of the creeks in the area.

The Mount Grainger field stretches down from the mountain itself back to Oodla Wirra, the township astride the Barrier Highway. Nackara Creek runs almost right through the township and northwards to the workings on the mountain. There are workings on either side of the creek along its entire length. This creek and its many tributaries provide first-class fossicking grounds with relatively easy access in many places, although landowners' permission will be needed.

Waukaringa and Teetulpa fields are reached by turning off the Barrier Highway at Yunta. While the road into Waukaringa is reasonable, the tracks leading off to the mines and creeks are rough and in many places can be negotiated only by four-wheel-drive vehicles. The Manna Hill field is fairly close to the highway, but here too the off-road tracks are rough. Access is northwards through the town of Mannahill on the Barrier Highway. Wadnaminga and New Luxembourg (Radium Hill) fields lie to the south of the highway, with much the same access as for the previous fields. The turn-off to Wadnaminga is at Olary, and access to Radium Hill is from a turn-off to the right some 25 kilometres to the east of that town.

There are other fields in this north-eastern region but most are too far

SOUTH AUSTRALIA

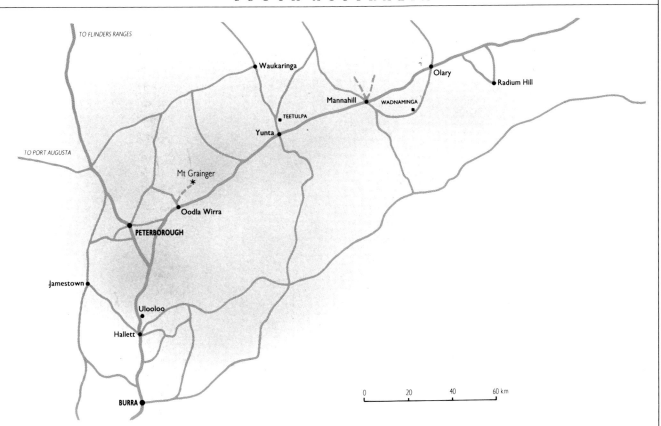

Map 21
The northern goldfields of South
Australia.

from the highway or too inaccessible for average family fossicking. Full
details can be obtained from the South Australian Department of Mines
and Energy in Adelaide.

Flinders Ranges
Like the ranges that sweep eastwards to Broken Hill, the Flinders Ranges,
running northwards from around Peterborough, are rich in minerals.
Mines of one kind or another will be encountered along the length of
these ranges, particularly in the north, between Lake Torrens and Lake
Frome, where a huge coal deposit is currently being worked near Leigh
Creek. Gold was found in this region during the gold-rush days and, as
is so often the case with old workings, fossickers still pick up enough
leftovers in the nearby Boolyeroo goldfield to make an expedition
worthwhile. A metal detector is once again an essential piece of equipment for successful fossicking in these dry areas.

The best road north to this area is via Peterborough or Port Augusta to
Hawker and then along the highway which follows the western slopes of
the ranges through Leigh Creek to Marree. Access into the hills and the
gold-bearing regions is through Hawker, Copley or Lyndhurst. Since
there are numerous tourist features in the Flinders Ranges, many access
roads are in good condition, but once off these roads the tracks may be
rough and require four-wheel-drive vehicles, particularly in wet
weather.

Angepena is probably the most accessible field, reached from a turn-off
at Copley. Bunyeroo is best approached through Wilpena, while the
northern fields lie mostly off roads that run out of Lyndhurst. The mines
in this area have mostly worked reef deposits, but a number of alluvial

73

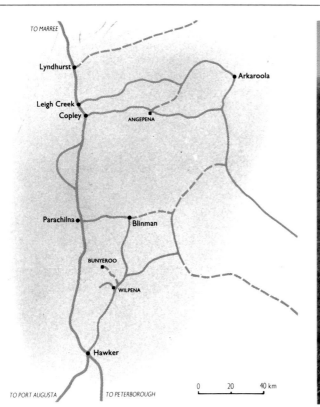

deposits have also been found. The main fossicking would be in the tailings of the mines and the creeks in the goldfield area. It must be kept in mind that this can be rugged country and care must be taken if prospecting far from the main tracks unless you have the right equipment and are used to this type of country. Water is scarce and although there are some arterial springs, their water is undrinkable.

Map 22
The Flinders Ranges region.

The rugged splendour of the Flinders Ranges was the scene of a number of early gold strikes. This is near Arkaroola.
PHOTO SHIRLEY HANKE.

Tarcoola area

Worked spasmodically until 1955, the mines in the Tarcoola area are among the most recent to produce gold in South Australia. However, only reef gold is found here, and this, together with the great distance involved in reaching the field, makes it an unattractive proposition for family fossickers. Tarcoola is about 200 kilometres beyond Woomera and located in uninteresting, inhospitable country where water is very scarce. Since there are numerous goldfields within 50 kilometres of the city of Adelaide, most of which are located in lush, rolling green countryside, it would be hard to imagine why any amateur would wish to endure the privations of fossicking the Tarcoola area.

WESTERN AUSTRALIA

Western Australia is one of the richest Australian states in terms of natural resources. Enormous deposits of iron ore are located in the north-west, oil and gas are found in a number of areas, both onshore and offshore, and even its vegetation has proved valuable, with the timber industry of the south-west playing a major part in establishing Western

Australia as a trading state. But, as is so often the case, it is gold that receives most attention, and today Western Australia has the largest output of gold of any state on the continent. In 1984 (the last year for which figures are available), 32,111 kilograms of gold were produced in Western Australia, 11,385 kilograms of which came from the East Coolgardie goldfield, centred on Coolgardie in the south of the state.

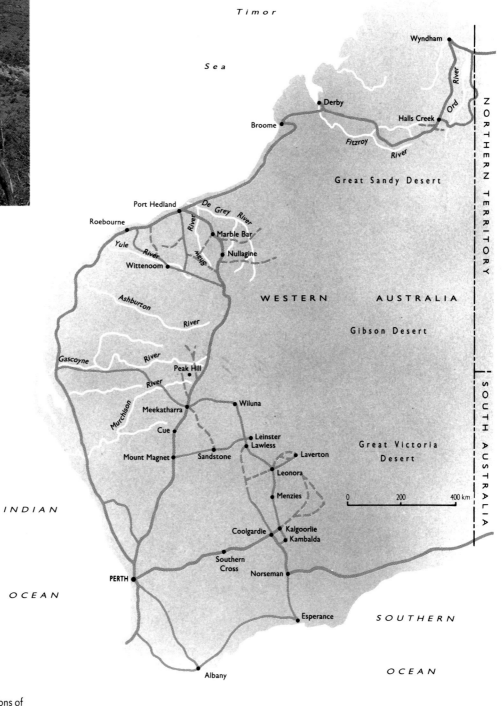

Map 23
The principal gold-bearing regions of
Western Australia.

Unlike most gold-producing states, Western Australia has maintained production over almost a century. When the eastern goldfields flourished, reached their peak, then collapsed, the deposits in the western regions maintained a lower, but nevertheless constant, flow of gold to the markets. The peak production of around 65 tonnes, recorded in the first decade of this century, has never been equalled, but despite downward trends elsewhere, production reached another peak, this time around 35 tonnes, during the early 1940s. In 1984 it had reached that figure again and was continuing an upward trend. Currently, Western Australia produces around 80 per cent of Australia's gold.

Although such figures would indicate that Western Australia is a gold fossicker's paradise, this is not quite the case. Even though there is a great deal of gold still to be won in the west, winning it is generally much harder for amateurs than it is in the eastern states. This is due to the nature of the gold itself and also to the climatic conditions of the country in which it is located. While much of the gold in New South Wales, Victoria and Queensland has been alluvial and won from creeks and rivers located in relatively pleasant countryside, usually with access to water, the conditions in Western Australia are mostly the reverse.

The main goldfields are well inland or in the northern regions where the countryside is dry and barren and water is at a premium. Creeks and rivers are dry for much of the year and recovering gold from the dry soil or gravel must be done with either metal detectors or the dry blowing process. Conditions are rough, hard, dusty and not at all conducive to pleasant family holidays. There are exceptions, of course, but generally,

Creeks and vegetation, including prolific grass-trees, make fossicking in regions near the coast less arduous than in the east.

prospecting in the western part of the continent is plain hard work with little appeal for amateurs.

The gold deposits are located in three distinct zones. The major gold-bearing rocks lie in a huge region of the south-west known as the Yilgarn Block. This covers the southern half of the state from just inland from the west coast to Laverton, on the edge of the Great Victoria Desert, and reaches from Norseman northwards to beyond Wiluna. Most of the gold deposits lie in the inland areas of the Yilgarn Block, particularly around the Kalgoorlie district.

A sizeable concentration of gold-bearing rock also occurs in the second zone, know as the Pilbara region, centred on Port Hedland, on the north-west coast. This is a much smaller field but has produced considerable gold since it was first mined. In 1984, about 4079 kilograms were extracted from Pilbara. An even smaller deposit, but one that created a significant rush when it was first discovered, is in the Kimberley region, centred on Halls Creek, in the far north of the state. Both these areas are fairly isolated. Wyndham and Kununurra, the only towns of any consequence near the Halls Creek goldfields, are some 2700 kilometres by air from Perth.

The Yilgarn Block

The major gold centres in this region are stretched across a wide section of the state. Those nearest to Perth are at Boddington, Lake Grace and Ravensthorpe, in the south-west corner. The Eastern Highway leads to Southern Cross where there are substantial goldfields in the adjacent districts, notably at Marvel Loch and Mount Palmer. This highway continues east to the most prolific of all Western Australian gold regions — that around Kalgoorlie, where commercial mining continues to this day. Of the gold workings in the area, East Coolgardie is the most prolific, with North Coolgardie and North-east Coolgardie producing substantial but much smaller results. Some 170 kilometres to the south is the old goldfield of Norseman.

The road north from Kalgoorlie to Leonora is 237 kilometres of reasonable surface, but access to the fields around Leonora and nearby Laverton can be rough and dangerous. This is survival country, virtually on the edge of the Great Victoria Desert, and not recommended as prospecting territory for family fossickers. To the north again there are scattered mines or sites along the road to Wiluna, but these are also in very arid country and great care must be taken when leaving the main road in these areas.

Other inland goldfields of the Yilgarn Block can be reached along the Northern Highway. Mount Magnet is one of the first and there are scattered fields along this highway past Meekatharra as far north as Peak Hill. Once again, the countryside is very arid and care must be taken when driving away from the main highway. Since there are few settlements, all supplies and requirements must be carried in, and once off the main roads, a four-wheel-drive vehicle is essential.

Pilbara region

The principal fields here are centred around Marble Bar and Nullagine, both of which are located on the Northern Highway. However, the

Once off the highway, the going can get rough in outback Western Australia.
PHOTO PHIL COLEMAN.

goldfields are scattered in the dry, rugged terrain around the towns and there are few access roads other than tracks suitable only for four-wheel-drive vehicles. Other fields located in the area across to Karratha suffer the same problem: few access roads and difficult, even dangerous conditions for those who are not experienced in this type of country.

A scene on the Fortescue River, in the Pilbara region. Most of the major gold finds in the north-west were located between this river and the De Grey River.
PHOTO TOM BUDDEN.

Typical gold country in the Kimberleys.
PHOTO PHIL COLEMAN.

The red rocks of the Kimberleys make a spectacular backdrop to the gold country in the Halls Creek area.
PHOTO PHIL COLEMAN.

Kimberley region

Like the Pilbara goldfields, those in the Kimberley region are located in rugged country with access only for four-wheel-drive vehicles and then only with experienced drivers. The Northern Highway runs right through the settlement at Halls Creek, but the roads from here out to the sites at Ruby Creek, Mount Dockerell and Grants Creek are rough. This region is subject to extreme weather conditions, particularly during the wet season, and care must be taken when working near rivers and creeks at this time of year.

NORTHERN TERRITORY

Take a map of Australia. Look for the largest expanse of blank space and you have found the Northern Territory. An overall view of the continent from a satellite in orbit would reveal that the blank space is not confined just to cartography but exists as a geographical phenomenon as well. The entire 1,346,200 square kilometres contains effectively one highway, one city and one town of any consequence. The rest is...space.

The ugly duckling of the Australian states, the "Territory", as it is known, has long been considered a hot, dusty, backyard region of interest only to cattle barons, miners and Aborigines. But the ugly duckling is changing. The tarnished down is giving way to pure white feathers, and the farmhouse in Canberra is slowly becoming aware that the region it has for so long treated so shabbily is emerging as a beautiful white swan.

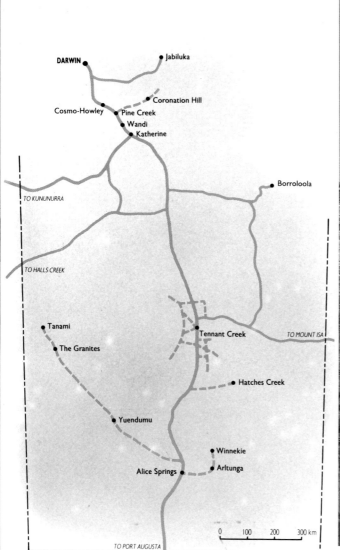

Among the blank spaces, some of the finest scenery on the continent is drawing thousands of tourists from around the world. Australians themselves are gradually awakening to the magnificent asset they have in their backyard, both above and below the ground. The Territory, which was never considered to have reached statehood, is ready to give the other states a run for their money. Where before only a few tough cattlemen headed north along the Stuart Highway, now tourist buses, camping expeditions and family cars flock across the South Australian border. The Northern Territory is now the "in" place for tourists, travellers and prospectors.

Prospecting has always been a feature of life in the northern regions, although extremes of climate and geographical conditions in the mineral-rich areas make fossicking a tough and difficult exercise. Distances are vast, much of the region is impenetrable other than by special vehicle, and danger lies around every sandhill or rock outcrop for the uninitiated and inexperienced. Some areas are total desert, many are inaccessible and waterholes are scarce. Large areas are designated as Aboriginal reserves, into which Europeans must not stray.

But despite this, some good finds have been made in the dusty red soil,

Map 24
Major goldfields of the Northern Territory.

Rich in all forms of minerals, the spectacular and in some cases unexplored ranges of the Northern Territory are possibly the last frontier for gold discoveries in Australia.
PHOTO ROBYN HILL.

not least of which was the phenomenal gold strike made in October 1987 by Ray Hall of Tennant Creek, who unearthed what could be one of the biggest lodes in Australia's mining history. His discovery gives hope and incentive to all prospectors, professional or amateur, who remain convinced that there are still great treasures of gold to be discovered beneath the dusty soil.

The earliest gold discoveries in the Northern Territory were made in 1865, and as prospectors flooded north, both to these new strikes and the discoveries in the adjacent areas of Western Australia, new deposits were found and new fields opened up. The more important of these areas were Hayes Creek, Pine Creek, Arltunga, Harts Range, Tanami and The Granites. Long after these initial rushes, a second wave of gold fever hit the Territory with the discovery of gold at Tennant Creek in 1933. These deposits were substantial and major Australian mining companies developed the Tennant Creek mines during the 1940s and early 1950s.

As with so many old fields, new technology and the high price of gold have given some of the early Territory mines a new lease of life. Apart from an increased interest in exploration in all areas, new developments, including open-cut mines, have been commenced on the old sites at Pine Creek, The Granites and other abandoned mines. Indeed, Pine Creek and The Granites, together with Tennant Creek, are currently the producers of most of the Territory's gold.

NORTHERN TERRITORY GOLDFIELDS

Pine Creek
220 kilometres south of Darwin on the Stuart Highway near the township of Pine Creek

Mount Bonnie region
approximately 140 kilometres south-east of Darwin

Cosmo-Howley
165 kilometres south-east of Darwin on the Stuart Highway

Coronation Hill district
220 kilometres south-east of Darwin via Stuart Highway and Pine Creek. Located in the Alligator River valley

The Granites
500 kilometres north-west of Alice Springs

Tanami
100 kilometres beyond The Granites

Nobles Nob area
18 kilometres east of Tennant Creek. Other mines in the area include White Devil, Black Angel and Northern Star mines (all operating)

Warrego
45 kilometres west of Tennant Creek

Argo
4.2 kilometres south-east of Tennant Creek

OTC8
5 kilometres west of Tennant Creek

Arltunga
approximately 120 kilometres north-east of Alice Springs

Winnekie
approximately 160 kilometres north-east of Alice Springs

Kurrundi–Hatches Creek
approximately 160 kilometres south-east of Tennant Creek, turn-offs north or south of Devils Marbles

Gold can be found in a number of well-scattered areas of the Northern Territory. Probably the most important, and the most accessible, is around Tennant Creek, a mining town on the Stuart Highway 536 kilometres north of Alice Springs. There are also deposits to the east and south-east of Darwin between the mining centres of Katherine and Jabiru and to the north-east of Alice Springs. Other areas, which have very difficult access, are the Barkly Tablelands, to the north-east of Tennant Creek and The Granites, in the Tanami Desert.

Fossicking is permitted anywhere in the Territory other than on Aboriginal land, mining leases, mining reserves and national parks. A Miner's Right is required and this can be obtained from any Department of Mines and Energy office on payment of a $5 fee. Since there are few areas where water is available, almost all fossicking in the Northern Territory involves the use of a metal detector or dry blower. Four-wheel-drive vehicles are essential and any foray into country away from centres of civilisation requires extreme care. Apart from ensuring that you have all the requirements both for the expedition and for survival if anything goes wrong, it is essential to advise the local police of intentions to "go bush" in this part of the country.

Although fossicking can be undertaken anywhere outside the restricted areas, the chances of striking gold are obviously better in the vicinity of a known deposit, or where mining was once carried out. The list of goldfields, past and present, on page 81 will indicate the location of known deposits, although care must be taken not to infringe current mining leases in the area.

Other than the main highways, roads in the Territory can be rough, and fossicking often requires four-wheel-drive vehicles and special care.
PHOTO MARIE AND BARRY AYERS.

TASMANIA

Tasmania, despite its geographical appearance, has not proved rich in gold. The mountainous regions of the north-east are the only areas of the island that have produced any worthwhile quantities of gold. It is of course more than possible that the relatively unexplored regions of the south-west have hidden deposits of the precious metal in the tortured rocks of wilderness areas. But if this is the case they will remain undiscovered for many years to come, perhaps forever, for this area is now under the umbrella of the world heritage system, which does not allow exploration, or for that matter even fossicking, in declared areas.

So Tasmania's known gold deposits are easily located, mostly lying in a belt that runs diagonally across the north-eastern tip of the island in a south-easterly direction from Waterhouse Point to the South Esk River near Fingal. Access to this region varies from main roads at either end, and the Tasman Highway running through Branxholm, to extremely rough roads in the mountainous regions between Mangana and Ringarooma. Gold-producing districts along the belt are near the centres of Waterhouse, Forester, Warrentinna, Alberton, Mount Victoria, Mathinna, Tower Hill and Mangana.

A few other mines recovered gold in other parts of Tasmania, notably in the Beaconsfield and Lefroy districts on either side of the Tamar River mouth. But in most other places the amount produced was insignificant or the field too isolated to be of interest to amateur fossickers. For the more intrepid, the relatively untouched areas of the west coast provide a challenge, for alluvial gold has been recovered from a few west-coast streams, such as the Savage River, near Corinna, and in the lower reaches of the Urquhart and Mainwaring Rivers.

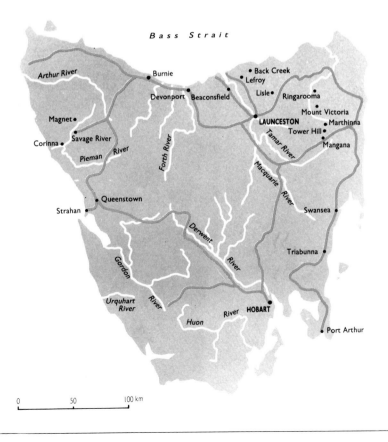

Map 25
Most of Tasmania's goldfields are concentrated in the rugged mountain regions of the north-east.

CHAPTER FIVE

SPARKLING TREASURES
FROM MOUNTAIN STREAMS

Of all the hidden treasures of Australia, probably the easiest to find and recover are the wonderfully coloured gemstones. Ranging from precious stones such as diamonds, to startlingly beautiful but less valuable thundereggs, gemstones in Australia come in a variety of forms. Some are of great industrial value because they are very hard, some have value because they are rare. But the appeal of gemstones lies mostly in their beauty — the magnificent ice fire of the diamond, the blood red of the ruby, and the kaleidoscopic colours of the opal.

There are so many varieties that fossicking for gemstones depends mainly on the type of stone required. In New South Wales, for example, almost every variety of gemstone can be found in areas with relatively easy access from a highway or major town — diamonds, opals, sapphires, topaz, beryl and emerald, to mention but a few. While there may be a preponderance of some gemstones in certain regions across the nation, most states have a worthwhile selection of gemfields within striking distance of their capital cities. This brings gem fossicking within reach of almost every enthusiast in the country. Since there are generally few restrictions imposed on fossickers, other than in commercial areas and national parks, it is a pastime that can be enjoyed with few

Amber, sphalerite, quartz amethyst and topaz.
PHOTO SCOTT CAMERON, COURTESY NSW DEPARTMENT OF MINERAL RESOURCES.

Opposite: The translucent beauty of precious gemstones creates an attraction for treasure hunters that is almost impossible to resist.
PHOTO DAVID BARNES, COURTESY NSW DEPARTMENT OF MINERAL RESOURCES.

inhibitions. It is relatively inexpensive, and since it takes place entirely outdoors, usually deep in the countryside, it is a healthy and wonderfully relaxing activity.

Gemstones fall into two broad categories — precious and semi-precious stones. Nowadays the tendency is to refer to them as gem minerals and ornamental minerals which, although perhaps a more accurate description, does not have the romantic flavour of the older terms. Gem minerals are the valuable stones, usually completely transparent, with superb colour and sparkle. Ornamental minerals are just as colourful, but usually translucent or opaque and lacking the sparkle of gems, relying more on attractive patterns to enchance their appearance.

Commercial operations are responsible for most of the gem minerals recovered. Australia accounts for almost the entire opal production of the world and over half the world's sapphires. In recent years, large-scale mining operations have opened up diamond deposits in the Kimberley region on a scale not seen anywhere in the world except in South Africa. Gemstones have always been popular in the fashion and jewellery trades, while the developing tourist industry is creating a rapidly increasing market, especially for opal.

Despite this, there is no shortage of gemstones for amateur fossickers. Most of the commercially recovered minerals are beyond the reach of amateurs, requiring deep-mining techniques or the use of heavy machinery to unearth them. But such operations can be advantageous for the less ambitious gem hunters. In the first instance, they locate hidden deposits which require sophisticated equipment or expensive exploration techniques, and would probably never have been found by amateurs. And secondly, their mining operations frequently leave "crumbs" which, although too insignificant for commercial use, send thrills of joy through a fossicker's heart!

As mentioned in the section on gold, working the tailings of an old mine or dredging operation can be profitable in these days of elevated prices, and the same applies to gemstones. Many a shrewd fossicking team works the streams near commercial mining operations for the leftovers, since the presence of a commercial outfit indicates that payable gemstone deposits are to be found in the area.

Sieving through the gravel of a softly rippling creek in the total tranquillity of the deep countryside, is a calming and restful therapy.

LOOKING FOR GEMSTONES

Originally formed in a wide variety of rocks, gemstones may be found in greatly differing environments. Some will be found in streams, some in igneous rock and some in sedimentary rock. As a rule, removing gemstones from the rock in which they crystallised is a difficult process without suitable machinery and likely to damage the gemstone. Most amateurs look for them in streams or stream beds, where nature has done the hard work, weathering the host rock into small pieces and even removing most of the rock that adheres to the gem. This makes recovery much easier and reduces the risk of damage when cleaning and polishing.

Gemstones are minerals usually formed in the extreme heat of volcanic upheaval and deposited in small quantities in cracks, joints and cavities of crustal rocks. As with gold, deposits close to the surface become weathered as the host rock is exposed to the elements, and break off as the rock disintegrates. Since gemstones are generally hard and durable, they withstand both the weathering process and the abrasion of being carried along in streams or rivers until they are deposited in a crevice or lodged behind an obstruction, where they settle with the detritus, mostly now reduced to sand and gravel.

It follows, then, that the places to look for gemstones are somewhat similar to those that produce gold. Not being quite as heavy as gold, gemstones will often travel farther in the stream and be less deeply buried than the precious metal; nevertheless, the locations will be similar. Sand or gravel banks on the inside bends of the stream are favourite spots, as are the bars that form where a stream joins a river.

But not all gemstones are found in streams. Opal is too soft to withstand the rigours of weathering and the abrasion of travelling down streams, while beryl (emerald and aquamarine) is too brittle. These minerals must be mined from the rock in which they are found, a tough and risky business with gemstones that are easily damaged. Fortunately for fossickers who have neither the time nor the experience to handle such demanding work, most of the popular gems can be taken from gravel beds in the streams of known gem-bearing areas.

A typical gemstone site in the New England ranges of New South Wales. Washdirt and gravel, carried down from the weathered rocks of the mountains, bear the fragmented gem minerals, which are deposited in pools and hollows.

FOSSICKING FOR GEMSTONES

There is no great difference between fossicking for gold and fossicking for gemstones. In both cases, there are two options — mining or working streams. Most amateur fossickers prefer panning. Because of the similar ways in which stream deposits of gold and gemstones are laid, the location of sites is much the same and there is little difference in the equipment used. Indeed, many gemstone deposits were discovered by prospectors looking for gold, and it is not at all uncommon to find both the precious metal and gem mineral in the same alluvial deposit.

There are a few small differences in the equipment used for some types of gem fossicking, depending on the location and the nature of the surrounding area. When working streams for gemstones, the gold panning dish is mostly replaced by, or used in conjunction with, sieves of varying gauges.

Since the gemstone habitat is more likely to be gravel and broken rock than sand or clay, the sieves are used to sift through the detritus, examining each grade of rock or gravel from the largest to the smallest. Many experienced fossickers, however, still use the dish for "puddling" stones from clay, mud, dust or sand. Obviously, the grade of sieve depends on the composition of the host pay-dirt or gravel, which also determines whether or not the dish is required. Sieves of about 6 millimetres and 3 millimetres mesh and about 350 millimetres diameter are best for most work, although a larger mesh might be required for the first sieving. Prospectors' stores and lapidary shops supply a range of sieves, usually made of light-weight metal and constructed so that they fit snugly into one another for convenience. Fossickers working reef mineral will need a couple of sledgehammers, a miner's pick, a geologist's hammer and cold chisels to break up the rock.

Sapphire fossicking is an activity for all the family. It costs little, yet provides hours of interesting activity and often surprising rewards.

Like panning with the dish, sieving stream gravel requires a little experience and practice. The larger mesh sieve is used first, usually with the next grade of sieve beneath it. A couple of shovelfuls of gravel is loaded into the top sieve and worked by rocking the sieve first forwards and backwards, then from side to side with a firm action. This causes the heavier gem mineral to work its way into the centre of the sieve, making it easier to spot and recover. The sieved gravel, which has fallen into the next grade of sieve, is then similarly treated, using a panning dish to catch the fine residue that passes through. By this time, any sizeable gems will have been extracted from the sieves, and it remains only to pan the residue in stream water to detect any small fragments that may have slipped through. Experienced fossickers often lock the two sieves together so that both are worked with the one rocking action. However, this can become tough on the arms after a long session.

THE TREASURES TO BE FOUND

There are over 2500 minerals to be found in the rocks and streams of Australia. Almost 100 are recognised as having a quality which puts them into the category of a gemstone. Most are valued for their beauty, which relates to the stone's colour, lustre, transparency, brilliance and "fire".

Gem minerals are the attractive, transparent stones such as diamonds, sapphires, rubies and emeralds, and have universal appeal as valuable, "precious" stones. Ornamental minerals such as jasper, the agates, turquoise, chrysoprase, malachite and rhodonite, lack the transparency of gem minerals and depend on their colour and pattern for their popularity, which can wax and wane according to fashion and availability. Gems are sold according to their carat weight (1 carat = 200 milligrams) and are usually cut and polished to enhance their unique features.

Considered inferior by some purists, ornamental minerals such as agate are magnificently coloured and patterned stones.
PHOTO DAVID BARNES, COURTESY NSW DEPARTMENT OF MINERAL RESOURCES.

The brilliant colour of a gemstone is not due to pigmentation. It is achieved by the absorption of certain of the spectrum colours of white light by the silica structure of the gem. This structure has the ability to absorb certain wavelengths of light. The colour of the stone as seen by the eye is made up of the spectrum colours which have not been absorbed. If no wavelength is absorbed, the gemstone will appear white. If all wavelengths are absorbed, it will be black. The magnificent play of colour in opal is the result of diffraction of light from the packed silica spheres of which the mineral is composed.

Elements within the stone determine which colours will be absorbed. If it is a major element, the gem will always be the same colour. If a number of elements are present, the gem may occur in many different colours. A good illustration of this is corundum, which in its pure form is colourless. When it contains traces of chromium it is the beautiful red ruby. The same mineral with traces of iron and titanium forms the much sought-after blue sapphire.

WHERE TO FIND GEMSTONES

Gemstones are found across the length and breadth of Australia. But because they occur in different types of rocks, some are found in closely defined locations. Opal, for example, being located in sedimentary rock, will rarely be found in rugged granite mountains. Nor are sapphires likely to be found in the sandy basins of the inland regions. There are, however, many gemstones which are common to a number of locations. To provide a guide to the major areas where gemstones are to be found, it is necessary to deal with each mineral individually.

SAPPHIRE

One of the most popular gemstones among fossickers is the magnificently coloured sapphire. Although generally thought of as being transparent and in varying shades of blue, sapphires can in fact be completely colourless. They can also be shades of yellow, gold, pink, green, violet and brown, or two-toned, incorporating both green and blue, or blue and yellow. Sapphires can also be a beautiful deep red, when they are called rubies. While sapphires are to be found in many parts of Australia, ruby is a fairly rare gem mineral.

The mineral which forms sapphire and ruby is corundum, normally transparent and colourless but transformed into magnificent colours by trace elements. The greater the amount of trace element in the corundum, the richer and deeper the colour of the gemstone, and it is this degree of colour that affects the value of the stone. Since they were formed millions of years ago by the violent upheaval of volcanic activity, sapphires are mostly found in mountainous country and in the streams and rivers that wash down the detritus of weathered rocks. Like most minerals, some sapphires have been buried in alluvial deposits with the passing of time. These are located in deep leads and many have been found during gold or tin mining operations.

The main concentrations of sapphires are in the New England Ranges in New South Wales and near Anakie in Queensland. In both these areas

Second only to diamonds as the most prized of all gem minerals, sapphires come in a surprising range of colours.
PHOTO DAVID BARNES, COURTESY NSW DEPARTMENT OF MINERAL RESOURCES. CUT SAPPHIRES COURTESY G & J GEMS, SYDNEY.

Far right: A typical sapphire-bearing stream near Inverell.

commercial mining is carried out, and most land in the vicinity is covered by mining leases. However, the downstream reaches or adjacent streams which are not covered by leases usually provide fossicking potential for amateurs. As a general rule, only commercial operations work the deep leads and alluvial dirt buried beneath overburden. Amateur fossickers usually find sufficient reward in sieving and panning the wash-dirt of the local streams or the tailings of previous mining activities.

Principal fossicking areas for sapphires are:

NEW SOUTH WALES

New England region

The New England region north of Armidale and west of Glen Innes has the largest deposit of sapphires in New South Wales, possibly in Australia. It is an ideal spot for amateur fossickers, since it is high in the New England Ranges — a high-rainfall section of the Great Dividing Range — and has a wealth of streams, many of which carry sapphire. Indeed, it is probably true to say that in the Glen Innes–Inverell region virtually every stream carries sapphire. Although many areas are mined commercially, there is plenty of scope for amateurs and a number of fossicking areas are located in the region.

The accompanying map indicates known deposits of sapphire, most of which occur in the streams to the north of the Gwydir Highway. Other prolific areas are to the east of Glencoe and to the south-east of Inverell. Sapphire is also to be found, albeit not so prolifically, in streams to the south of the main field near Guyra and Armidale. Outside the New England district, sapphire occurs in a number of areas throughout New South Wales. Those mentioned here have revealed surface deposits in streams and creeks.

Barrington Tops

Blue sapphires, rubies and mauve-coloured corundum have been found in the streams and creeks that drain this mountainous area. The stones are of good quality.

Berrima–Mittagong

Sapphires and other gemstones have been found in a number of areas around Mittagong and Berrima, mostly in alluvial deposits in the Wingecarribee River. Stones up to 20 millimetres across have been discovered in these deposits, although many have been opaque and of low quality. Streams in the vicinity of nearby Robertson district have also produced sapphire, mostly a green variety.

Crookwell district

Sapphires are found in the alluvial deposits of Grabben Gullen Creek, Wattle Creek, Wheeo Creek and other streams in the area to the south and west of Crookwell. Most are deep blue, but many are patchy in colour, favouring the mauve and purple regions of the spectrum. Large, opaque blue sapphires have also been found in this area, while some unusual black sapphire has been recovered to the east of Crookwell.

Hill End

Sapphire and other gem minerals are to be found in streams in this area, which is also a popular gold fossicking location.

Kiandra

Although most of the sapphires found in this area were discovered during the goldmining era, there is every possibility of further deposits occurring in the local mountain streams. The sapphires taken from Kiandra were green.

Macquarie River

Because it drains the western side of the Great Dividing Range, an area rich with mineral resources, the Macquarie River has good potential for alluvial minerals and gold. Sapphires and other gemstones can be taken from alluvial deposits in a number of places, notably in the vicinity of Wellington and near the Burrendong Dam. The first sapphires discovered in New South Wales were found in the nearby Cudgegong River, close to Gulgong.

Mount Werong

The creeks surrounding the Mount Werong plateau have produced many sapphires although most have been small. The alluvial deposits of Lanigans Creek, Limeburners Creek and Werong (Ruby) Creek have been particularly prolific. The stone is mostly light blue or green.

Nimmitabel region

Sapphires have been found in the alluvial deposits of the Maclaughlin River, about 24 kilometres south-west of Nimmitabel, and also from streams near Bemboka. The stones were blue and fairly small.

Nundle

Creeks to the north-west of Hanging Rock have significant deposits of

1. Bingara
2. Inverell–Glen Innes
3. Narrabri
4. Copeton
5. Oban
6. Bendameer
7. Nundle
8. Dunedoo
9. Barrington Tops–Gloucester
10. Mudgee
11. Gulgong
12. Hill End
13. Airley Mountain
14. Oberon
15. Mount McDonald
16. Mt Werong
17. Crookwell
18. Berrima
19. Wee Jasper
20. Mittagong
21. Kiandra
22. Tumbarumba
23. Nimmitabel

Map 26
Principal locations of sapphires and other gem minerals in New South Wales.

sapphire and other gemstones, particularly large brown zircon. The sapphire is usually dark blue and often opaque.

Oberon

Many streams between Oberon and Porters Retreat are known to carry sapphire. Native Dog Creek, close to its headwaters, first revealed the presence of the gem in the area. Since then alluvial deposits of sapphire have been found in Campbells River between Mount David and Daisy Bank, the Isabella River near Isabella, the headwaters of the Fish River, the Retreat River near the Oberon–Taralga road, the Abercrombie River near Mount McDonald, streams near Black Springs, and Grove Creek near the Abercrombie Caves.

Native Dog Creek, a tributary of the Campbell River, one of the most prolific waterways in the Oberon district.

Shoalhaven River

A few sapphires have been found in this river by prospectors panning for gold.

Tumbarumba district

Ruby Creek and Reedy Flat are the main locations for sapphire in this area, but many of the streams in the surrounding area are believed to carry the gems in their alluvial deposits.

QUEENSLAND

Anakie

Just as New South Wales has a concentration of sapphire-bearing country in the northern New England district, so the main Queensland deposit of this beautiful gemstone is concentrated in one area. About 320 kilometres west of Rockhampton, along the Capricorn Highway, is an area where local name places leave no doubt as to the reason for their existence. Emerald, Sapphire and Rubyvale are a treasure-seeker's paradise, with extensive gemstone deposits in a number of areas, the most important being the Anakie sapphire field.

This prolific region, which is centred on Rubyvale and Sapphire, now has the highest output of sapphires of any commercial field in Australia, and accounts for 80 per cent of the world's production. It is renowned not only for its superb blue sapphires but also the rich golden sapphires of the Willows field, which, in 1952, produced a golden sapphire weighing 91 carats. Although commercial mining is still in progress, the area is mainly the focus of fossickers and amateur miners. In 1987 a promotional campaign run by the Queensland Government won increasing numbers of fossickers to the region and the area is rapidly becoming a major tourist attraction.

From Emerald, access to the fields is westwards along the Capricorn Highway for 42 kilometres to Anakie, where a turn-off leads directly to Sapphire and on to Rubyvale, in the very heart of the main gemstone area. From either of these centres access is available to many of the designated mining areas. A further 20 kilometres along the Capricorn Highway another dirt road leads into the Glenalva designated area, while 7 kilometres further on again is the turn-off to the famous Willows gemfield.

For the most part, access to the Anakie fields is reasonably easy for conventional vehicles in fine weather, though some of the outer fields require the use of four-wheel-drive vehicles. However, because all roads other than the highway are dirt and many just tracks, heavy rain can cause problems, particularly for two-wheel-drive vehicles. From any of the main centres, and in particular from Sapphire and Rubyvale, a designated field or fossicking area is only a short distance away, and most places have good facilities, including caravan parks, camping grounds and motels. Excellent material for amateur gem hunters, including maps and a fossicking kit, are available from the Queensland Department of Mines, 61 Mary Street, Brisbane, and local information can be obtained from the local tourist centre.

Other districts

Although useful quantities of sapphire have been recovered in the Stanthorpe and Herberton districts, most finds in other areas have been either insignificant or located in such isolated regions as to be virtually inaccessible to amateur fossickers. With access so easy at the Anakie fields, and a good concentration of gem mineral providing almost certain success, there would seem little point in amateur prospectors going elsewhere, particularly into difficult country where there is no guarantee of finding gems.

Holding the sapphire against the light reveals its full translucence. Generally speaking, the more transparent the gem, the greater its value.
PHOTO DAVID BARNES, COURTESY NSW DEPARTMENT OF MINERAL RESOURCES.

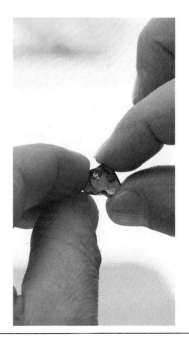

Right: Recovering sapphires at the Big Bessie field near Anakie. The paydirt is first sieved to separate large stones and rocks from the gem-bearing soil.
PHOTO VICKI ARMSTRONG.

Washing the gem-bearing dirt. The sieve is dunked rapidly in and out of a drum of water, then inverted quickly onto a flat surface. The heavy gems, having sunk to the bottom of the sieve, then appear on top of the washdirt.
PHOTO VICKI ARMSTRONG.

Coral

Sea

O'Briens Creek ●
Lava Plains ●
Agate Creek ●
Cheviot Hills ●

Binbee ●

QUEENSLAND

Marlborough ●
Anakie ● Mount Hay ●

0   150   300 km

Proston ●
Mapleton ●

Cedar Creek ●

Map 27
The principal gemstone fields in Queensland.

VICTORIA

Sapphire deposits are not as common or as concentrated in Victoria as they are in Queensland and New South Wales, although occasional finds have been made in streams washing down from the Great Dividing Range. Many have been discovered by fossickers panning the mountain streams for gold and a few localities away from the ranges have also produced sapphire in alluvial gold deposits.

Beechworth is a popular area for many gemstones, including sapphire, which has also been found in the Yarra Valley, the Dandenong Ranges, Rutherglen, Daylesford, Phillip Island and the Mornington Peninsula. A violet variety of corundum known as barkylite is found in the Ovens district.

SOUTH AUSTRALIA

Few significant deposits have been found in South Australia, the only sapphires of any consequence being located near Mount Painter in the north-eastern region of the Flinders Ranges.

TASMANIA

Many of the mountainous regions of Tasmania have a geographical resemblance to the New England area of New South Wales, and so might be expected to produce sapphires. Deposits do occur at Stanley River, a tributary of the Pieman River, on the west coast; at Table Cape on the north coast; and in a number of areas in the north-eastern region, particularly near the Weld River. But for the most part the number and quality of stones recovered does not encourage much commercial exploitation, nor great interest from amateurs, although gem quality stones of four to eight carats are found in the Weld River area. Sapphires are also located in alluvial deposits near Gladstone.

OTHER GEMSTONES

To cover the full range of gem minerals available in the creeks and gullies, the hills and ranges of this country would take volumes. There are many excellent publications on the subject, many of which are in the list of further reading at the end of this book. But since gemstone fossicking is only a part of the search for hidden treasure in Australia, it can be dealt with in a fairly superficial way here. Nevertheless, since many readers of this book will be looking at treasure hunting for the first time, a basic introduction is necessary to get the family motivated and out into the field. When their efforts are rewarded by a magnificent sapphire or topaz, perhaps then will be the time to look for more technical and in-depth information on the subject.

Fossicking out information on the whereabouts of a potentially rewarding site can often be as much fun as fossicking the site itself. Most capital cities house state mining authorities who are generally very helpful to gemstone hunters, and many locations near gemfields, such as Rubyvale in Queensland, have stores and mining and historical museums that contain a wealth of information reduced to a non-technical level that can be understood by amateurs. It is not uncommon to find as

Lesser known, but equally beautiful, fossilised wood is a much-treasured ornamental mineral.

PHOTO DAVID BARNES, COURTESY NSW DEPARTMENT OF MINERAL RESOURCES.

many treasure hunters in these places as in the creeks and gullies of the gemfields!

Australia's wide range of gemstones covers every colour in the spectrum, and knowing and identifying what appears to be a worthwhile stone is an important — and exciting — part of the treasure hunt. Chips of quartz, passed over as commonplace before, start to take on a new interest, particularly if you spot a faint tint of colour or an unusual reflection. Because of the enormous range of gemstones, many treasures are bypassed because the finder did not appreciate their value. Research can add another dimension to the excitement of the hunt by revealing unknown facets of the fascinating subject of geology. What was perhaps dry and uninteresting at school becomes alive and totally absorbing when carried into practice tracking down gemstones.

The details provided here of the locations of known gemstone deposits cover only the more accessible and better-known gemfields, although literally any creek or stream in Australia has the potential to produce gemstones. For amateurs, the best chance of success lies in the established regions, and it is these that are listed below, dealing in turn with each of the better-known gems.

DIAMONDS

Although small amounts of diamonds have been found in most states at some time or another, the only commercially viable deposits were mined in New South Wales, and these only in limited quantities. Now huge

Seen from a distance, tailings from the massive Argyle diamond mine in the Kimberleys appear to be building a new range of mountains.
PHOTO PHIL COLEMAN.

quantities of high-grade diamonds are being recovered from the Kimberley region of Western Australia, placing this country among the world leaders in diamond production.

During the mid-nineteenth century, when the earth was torn apart in the frantic search for gold, diamonds were discovered in a number of places in New South Wales, notably at Suttons Bar, and at other points along the Macquarie River. A big strike was made in the Cudgegong River, near Gulgong, in 1867. While working a deep lead for gold, more than 3000 diamonds were recovered, giving rise to the first commercial diamond venture. But insufficient quantities were found to make the operation successful. Diamonds were again recovered from the Macquarie River in 1950.

The most significant find was in the alluvium of Copes Creek in 1875, when diamonds were recovered close to its junction with the Gwydir River. This led to the discovery, in 1883, of the most productive field in New South Wales. Commercial quantities were found in the deep leads at what is now known as Copeton, on the edge of the waters of Copeton Dam, near Inverell in the northern New England region. Good deposits were also discovered at nearby Bingara. During World War II, Mr Tom Heath, the only full-time diamond miner in Australia at the time, kept the Lithgow small arms factory supplied with industrial diamonds which he sluiced from the Copeton deposit.

Other areas in New South Wales which have produced diamonds are Mount Airly, near Lithgow, and a number of places in the Central Highlands, notably Mount Werong, where a 28 carat diamond was found in 1905. Some finds have been made in the nearby Abercrombie and Crookwell rivers, mostly in the course of working for gold. A commercial mine once worked one of the headwaters of the Nepean River near Mittagong. The mine was known as Southey's Diamond Mine, and the creek was Doudles Folly Creek. Only a few diamonds were recovered and the venture failed.

Discoveries of diamonds in any quantity are not recorded from states other than Western Australia, although a few have been found in Tasmania, near Corinna, and in Queensland, near Stanthorpe. The massive deposits in the Kimberley district of Western Australia have exceeded any previous diamond discoveries in the mining history of this country. There is no public access to the region where this mining is carried out, however, and it is unlikely that amateur fossickers will be permitted in the area for many decades to come.

TOPAZ

Usually pale blue or colourless, Australian topaz is found mostly in gold- and tin-bearing alluvium. Waterworn fragments are also found in streams. It is confined mostly to the eastern areas of Australia, in particular to the stream gravels of the East Australian Highlands. The Oban district of New South Wales, some 40 kilometres south-east of Glen Innes, is one of the main centres for recovering topaz. A fine stone of 184 carats was once retrieved from this area. The nearby Torrington district also has good deposits of topaz, which have been recovered from Blatherarm Creek, Scrubby Gully and in nearby Bald Rock Creek. Other areas in the northern New England regions are also known for their topaz, usually in tin-bearing alluvium. Tingha, Elsmore and Copeton are good areas to fossick, as is the Gulgong–Mudgee district, farther south and to the west of the ranges.

In Queensland, the Mount Surprise and Mount Garnet districts have both produced good topaz finds, while in Victoria, waterworn fragments have been found in alluvial deposits at Beechworth, Gembrook, Maldon and Dunolly. Blue transparent topaz has been recovered at Melville, in Western Australia, some 250 kilometres west of Geraldton, and at Grosmont and Londonderry, near Coolgardie. Killiecrankie Bay, on the western side of Flinders Island, Tasmania, is known for its waterworn pebbles of topaz in tin-bearing alluvium, and other finds have been made near Mount Cameron, in the north-east corner of the island.

Treasures of the streams. A variety of precious gemstones, together with a handful of blue sapphires, create a fine prize for an industrious fossicker.
PHOTO PAUL HANKE.

GARNET

Gem garnet is usually found as waterworn fragments in stream beds. The most common type is red and is often confused with pink sapphire or zircon. At one time, good garnet was passed off as ruby, which is a much rarer and more valuable gem. Garnet is mainly found in stream gravels, the best-known areas being the Hale and Maude Rivers in the Harts Range of the Northern Territory, and also near Proston west of Gympie in Queensland. It is also found in some quantity in metamorphic rocks in the Broken Hill area.

BERYL (EMERALD AND AQUAMARINE)

The beautiful deep green of emerald and the paler blues and greens of aquamarine are the two best-known forms of gem beryl. Of the two, emerald is the more attractive and valuable. Softer and more brittle than most gemstones, beryl is rarely found in sizeable pieces, and when waterworn it usually takes a cylindrical shape. It is often found in association with alluvial deposits, particularly of tin.

Deposits of emerald and aquamarine are widely scattered, but few are of commercial value. Mining operations on a commercial scale are carried out in a few areas, mostly for emerald, but both types of beryl are recovered more in association with other mining activities, such as tin, wolframite and cassiterite, than as the result of specific beryl mining.

In New South Wales, gem quality beryl is often taken during mining operations at Emmaville tin lodes and at Heffernan's Wolfram Mine, near

Torrington, both in the New England region. Emeralds, in association with cassiterite and topaz were mined near Emmaville, but despite the discovery of good quality gems, there was insufficient to maintain the operation.

Western Australia has the main commercial emerald fields in Australia, located in the arid central-west of the state at Poona, near Cue and near Menzies, to the south-east. A 138 carat emerald, the second largest in the world, was found at the Poona mine, which was once owned by the Aga Khan. Aquamarine is also found at these centres as well as at Melville and Wodgina. The only other beryl deposits of significance are of aquamarine and are found in the MacDonnell Ranges and the Harts Range in the Northern Territory. Intermittent mining has also recovered aquamarine from the Chillagoe, Herberton and Stanthorpe districts of Queensland.

Sorting gemstones back at camp is the highlight of a fossicking expedition.
PHOTO PAUL HANKE.

ZIRCON

This gemstone comes in a wide variety of colours, depending mainly on where it is found. The New England area of New South Wales is a major centre for this attractive, often valuable gemstone, although it can be found in streams in many parts of the country.

Cinnamon-coloured zircons are found in the basalt soil of the Hanging Rock area, while there are red waterworn fragments in the Rocky River near Uralla. Most streams and creeks on the highlands of the New England Range are sources of zircon, a brilliant golden-yellow variety being found in the Inverell district. The southern highlands of New South Wales is another area where zircon is found. Wherever sapphire is found, zircon is also likely to be in evidence as the two seem to go together.

The Daylesford–Hepburn area, the Mornington Peninsula and Phillip Island are prime locations for zircon in Victoria. The Northern Territory was once a popular place for zircon fossickers with its widespread deposits on the surface of the Strangway Range, some 80 kilometres north of Alice Springs. But after a mini-rush in 1940, when the deposit was first discovered, the range has been so well worked that there is little zircon of any consequence left. Queensland's Anakie sapphire fields are a natural source for zircon, and some fine gems have been found in this area.

Other notable deposits of zircon occur in Tasmania, where alluvial deposits at Sisters Creek and Boat Harbour (near Wyndham), at Penguin, and in the Weld River, have produced good results in the past. Like the New England ranges of New South Wales, the rugged mountain country in the north-eastern corner of the island state has widespread deposits of the gem, and prospecting any stream in the region could well produce rewarding results.

QUARTZ

Quartz is the most common of all minerals and comes in many different forms. Clear, transparent quartz is in itself an attractive gemstone which is used for a number of purposes, not least of which is "capping" an opal

Much of the attraction of gem hunting lies in the peaceful environment and the outdoor living.

The crystalline purity of colourless quartz gives the gem an ice-like quality.
PHOTO DAVID BARNES, COURTESY NSW DEPARTMENT OF MINERAL RESOURCES.

triplet — a thin veneer of quartz is fitted over the opal to protect it and to enhance the play of colour.

The yellow-brown variety known as citrine is extremely popular in jewellery and can be cut and faceted to resemble a much more expensive gem. Both this and the smoky-blue variety of citrine are found in the Oban region of the New England ranges of New South Wales. Smoky quartz is also popular as a decorative gem, ranging in colour from grey-brown almost to black, which when polished and faceted bears a strong resemblance to smoky topaz. This form of quartz is also found in the New England region as well as in Tasmania.

Probably the most popular and valuable gem quartz is the delicately tinted amethyst. Ranging mainly through shades of purple, this gem-stone is located principally in the New England ranges and in the Wave Hill district of the Northern Territory. Large amethyst crystals have also been found in Queensland, near Bowen, Collinsville and Murgon, while a few have been found in Victoria in Cardinia Creek to the east of the Dandenongs.

The north-eastern corner of Tasmania is a rewarding gem-fossicking area, producing good finds of quartz. Citrine and amethyst are to be found in the Blue Tier, Rossarden and Mount Cameron areas, as well as in the Little Swanport region, where finds of fine amethyst have been reported in the past. Much the same applies to the Hardey River area of Western Australia, where amethyst was taken in quantity by early prospectors.

The proliferation of gem-quality quartz throughout the whole of Australia makes it difficult to pinpoint any particular location as being a good fossicking spot. The mineral can be found in one form or another in almost any creek or stream across the nation, and the problem is not so much to locate deposits of quartz but to seek out those of finer quality.

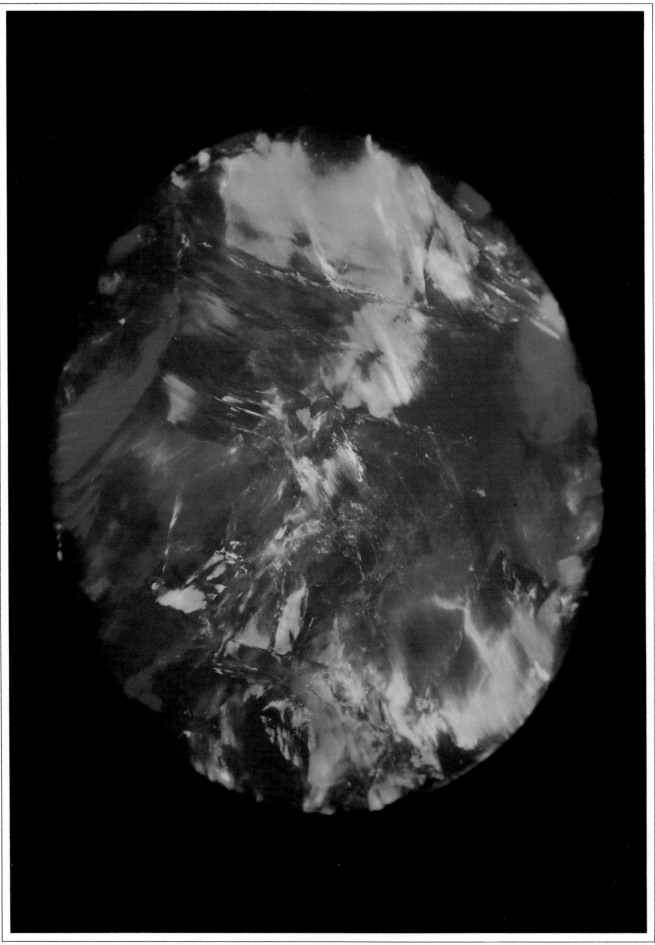

CHAPTER SIX

LIQUID COLOUR
FROM THE SOIL

No other gemstone has the magnificent variety of colour and brilliance that is found in opal. No two opals have exactly the same colour, and a single stone can contain the full range of colours in the visible spectrum. It is small wonder that these magnificent gems have been prized as rare and valuable stones almost since the beginning of time — the Romans are believed to have known about them as early as 250 BC. Unlike most gem minerals, which are the product of the fiery era when the earth's crust was molten, opal was formed in existing rocks by the seeping action of water containing a rich concentration of silica. The water permeated through cracks in the rock, usually sedimentary rock, collecting in pockets or seams, often next to a layer of impermeable rock. Once locked into one of these cavities, the water was gradually absorbed by the surrounding permeable rock, concentrating the silica into a gel which slowly hardened to form opal.

Most of Australia's major opal deposits have been formed in this way and lie mostly in the sedimentary rocks within the Great Australian Basin. These deposits are located underground, but at a relatively shallow depth — usually less than 30 metres beneath the surface. This, and the relatively soft host rock,

Opal from Mintabie.

PHOTO SCOTT CAMERON,
COURTESY NSW
DEPARTMENT OF MINERAL
RESOURCES.

Opposite: The liquid fire of gem opal is created by light reflected from millions of tiny spheres within the structure of the precious stone.

PHOTO DAVID BARNES,
COURTESY NSW
DEPARTMENT OF MINERAL
RESOURCES. OPALS
COURTESY SAPPHIRE AND
OPAL CENTRE.

makes opal mining an easier proposition than mining gemstones from reefs and seams contained in volcanic rocks. Opal mining is within the scope of amateur prospectors, although it requires experience, dedication, a certain amount of equipment and a lot of hard work.

Unlike fossicking through stream gravels or panning river silt, mining for opals is a very demanding occupation even on a small scale. To begin with, most of the major opal fields are in harsh, arid country where extreme temperatures, lack of water and absence of creature comforts make fossicking a far less attractive proposition than when carried out in pleasant mountain or bush country. The gem mineral is found only in its host rock, and so there are no easy ways of winning it; every piece must be physically hacked from the rock bed.

Offsetting these drawbacks, however, is the lure of the magnificent gem itself. Just to possess a fine piece of opal is rewarding. To win it from the rock where it has lain for perhaps millions of years, adds a compelling incentive that few can resist. Each year the lure of the multi-coloured stone draws thousands from the comfort of their homes out into the arid wastes of the inland, to labour under excruciating conditions.

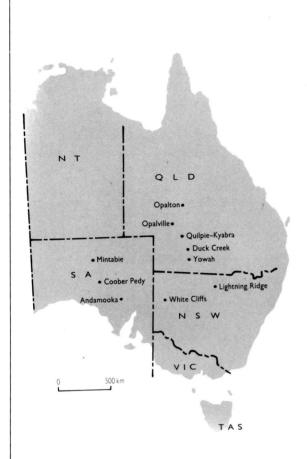

Opal embedded in the host rock where it formed perhaps millions of years ago. The milky, opaque type seen close to the finger is known as "potch".

PHOTO DAVID BARNES, COURTESY NSW DEPARTMENT OF MINERAL RESOURCES. OPALS COURTESY SAPPHIRE AND OPAL CENTRE.

Map 28
Location of Australia's principal opal deposits.

TYPES OF OPAL

While there are numerous different types of opal, most fall within one of two classifications: precious opal (also known as gem opal) and common opal. Precious opal is the only variety that has commercial value. The difference lies in the play of colour or fire in the opal. The gem is made up of millions of tiny spheres which break down white light into different colours, just as the prism of a spectroscope breaks down white light into the colours of the spectrum.

If the millions of tiny spheres in a piece of opal are arranged in a regular pattern, a brilliant play of colour is produced and the opal is classified as gem opal. If the spheres are arranged in a hotch-potch, or uneven pattern, then no play of colour is produced and the opal is termed common or potch opal. Both types are found in close association, with the gem opals sometimes located in bands within the potch.

The two main types of precious or gem opal are white opal, a translucent, milky-coloured gem found in a number of fields, particularly in South Australia, and the more sought-after black opal. This is mostly taken from the field at Lightning Ridge, where some of the finest specimens of black opal in the world have been found. It is dark grey or black, sometimes almost opaque, and throws a brilliant display of colours.

Other forms of precious opal are known as Harlequin, when it has a

mosaic or chequered pattern of colour, and Fire, when it reveals flashes of red. Pin opal shows closely spaced specks of brilliant colour, while Flash opal, as its name denotes, gives off flashes of colour. Milk opal has a white or cream body colour in which the opal colours can be seen, while Girasol has a bluish or reddish play of colours against a pale body colour.

Common opal or potch is not usually valuable because of its lack of colour compared to its precious counterpart. It is used commercially as backing for precious opal pieces which are thin or need darker opal to enhance their colour. When a slice of common opal is cemented behind a veneer of precious opal it creates what is known as a doublet. When such a combination is then protected by an overlay of clear quartz, it forms a triplet.

Despite its relatively low value, potch can be a rewarding find for amateurs where the pleasure of recovering a piece of opal is as important as any monetary gain. This is particularly true of fossil opal where the potch has replaced the bones or shell of a fossilised creature and the shape of the fossil can be seen in the potch.

There are many instances of opal, both precious and common, being found in fossil remains. The White Cliffs and South Australian fields are teeming with opalised fossil shells. One of the most important finds was made near Lightning Ridge in 1987 when opalised bones from two dinosaurs were unearthed by a team from the Australian Museum. One leg bone, composed mostly of grey potch, had two seams of precious opal through it and is estimated to be worth more than $20,000.

WHERE TO FIND OPAL

Although the major opal fields are in the inland areas where the gem has formed in sedimentary rocks, opal can also be found in smaller quantities in the volcanic rock of mountain ranges. Like many other gemstones, opal has been found by prospectors searching for gold and it has also been found in association with other gemstones in highland regions. Few of these finds have proved of commercial value, sometimes because of the limited extent of the deposit, sometimes because of the inferior quality of the opal. However, such places are ideal for fossickers and are mentioned in this section.

Although underground mining is the method generally used by small operators, much of the opal won in recent years, particularly in the Lightning Ridge fields, has been produced by reworking old mullock heaps. "Puddling" is a method of sieving opal dirt by mechanical means. The mullock dirt is thrown into a revolving perforated drum which breaks it up and discards the finer dirt, retaining the hard opal-bearing nobbies or individual nodules for closer examination. Noodling, which is the only form of opal fossicking carried out on a casual basis, simply means picking over the mullock heaps or tailings of previous mining activities. It goes without saying that opal miners guard their territories zealously and it is important to seek permission before noodling in a working area. Most miners are very co-operative with amateur noodlers, providing they take the trouble to ask first.

NEW SOUTH WALES

There are two major opal fields in New South Wales: Lightning Ridge in the north of the state, and White Cliffs in the north-west. Both have been commercially viable but only Lightning Ridge now produces commercial quantities of gem opal. Both fields lie close to major highways; Lightning Ridge is only 6 kilometres off the Castlereagh Highway and White Cliffs some 100 kilometres from Wilcannia and the Barrier Highway. A number of smaller but still important fields lie within a 50 kilometre radius of Lightning Ridge.

All New South Wales gem opal is won by mining, although often the mineral lies close to the surface. Because of this, either open-cut or underground methods are used. Most individual miners use a single shaft sunk by pick and shovel, with a windlass or winch to raise buckets of opal dirt or waste. When traces of opal are found, a level shaft or tunnel is excavated to follow the deposit.

Fossicking, or noodling, through the mullock heaps and tailings is the method used by most tourists (and many locals) to find their opals. The most productive waste heaps for noodling are those excavated by bulldozers at the open-cut mines, but many underground miners will also allow their mullock heaps to be turned over by noodlers.

LIGHTNING RIDGE

Apart from its size and extent, the Lightning Ridge opal field is famous for its unique black opal. The field is situated on the north-west plains of New South Wales, 76 kilometres out of Walgett and 770 kilometres from Sydney. There is a sealed road all the way, and with the recent increase in commercial opal mining in the areas, facilities have improved and a sizeable permanent population is established in the township.

The main area of this field forms a V shape about the north–south axis, each arm extending some 8 kilometres from a base near the Castlereagh Highway. The township is along the right arm of the V, as are most of the workings, which extend to the north, east and south of the town. There are four prescribed fossicking areas among these workings.

The opal mostly occurs in thin seams or layers in both vertical and horizontal joints of the rock, or as nobbies. The sediment that contains the gem is a light buff-coloured claystone known as opal dirt, which tends to harden and whiten when dry. Generally, layers of opal dirt are located between 6 metres and 18 metres below the surface, and because of this, the main method of recovering the gem is by mining. A shaft must be sunk through the surface layers of sandstone to reach the layers or lenses of opal dirt. A thin but hard layer of ironstone known locally as the steel band may be encountered before reaching the first opal dirt lens.

Major workings in the Lightning Ridge field are indicated on the map. A number of smaller fields are to be found mainly to the west, and since all of these have produced good results, they are described here:

Glengarry

To the south-west of Lightning Ridge and accessible over some 60 kilometres of dirt road, this field has produced good results over the years and was the scene of an opal rush in 1970 when a big find was made. In the six months between January and June 1971, an estimated $600,000

The moonscape of Lightning Ridge. Some of the world's finest black opals have come from this field.

PHOTO DAVID BARNES, COURTESY NSW DEPARTMENT OF MINERAL RESOURCES.

An opal "nobby" from Lightning Ridge.
The fire of the gem is seen quite clearly
against the lustreless grey of the host
claystone.

*PHOTO DAVID BARNES, COURTESY NSW
DEPARTMENT OF MINERAL RESOURCES. OPALS
COURTESY SAPPHIRE AND OPAL CENTRE.*

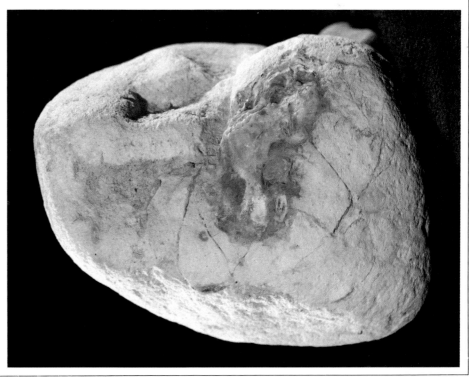

worth of opal was taken in this field. The opal dirt layers occur at depths of between 2.7 metres and 4.5 metres and are quite extensive.

Grawin

Close to Glengarry, this field is also fairly extensive and has been rewarding for miners. The opal is mainly in seams, some of which are only 1.2 metres below the surface.

Coocoran

This is an older field which has produced good results at times from opal dirt levels between 3.6 and 15 metres deep. It lies about 18 kilometres west of Lightning Ridge.

New Coocoran

In contrast to neighbouring Coocoran, this field was first opened up in 1972 and opals have been found very close to the surface.

Carter's Rush

Another new field, having first been mined in 1974. The returns have not been spectacular and the opal dirt lies fairly deep beneath the surface. This field lies about 5 kilometres north-east of Grawin.

New Angledool (Mehi)

This field, some 11 kilometres north-west of the small township of New Angledool, was first discovered in 1924. It produces opals with good colour, but mining was discontinued relatively soon because the gem was found to be brittle and did not cut well.

Sheepyard

A relatively small, new opal field, this has been the scene of some excitement since the discovery in 1987 of a number of opalised fossil bones. The field is about 70 kilometres west of Lightning Ridge and access is over a dirt road.

Other deposits

In a number of areas away from the established opal fields, opal occurs occasionally in cavities in volcanic rocks. Precious opal found in these places has mostly been brittle with a tendency to craze when taken from its host rock. For this reason, as well as the relatively small amounts of precious opal found, no commercial mining has taken place. The major locations are Tintenbar (near Ballina), Rocky Bridge Creek (south of Blayney), Mullumbimby, and Tooraweenah in the Warrumbungle Mountains.

WHITE CLIFFS

The only other major opal field in New South Wales, White Cliffs was the first commercial field to be worked in Australia, with the first shaft sunk around 1889. When it reached its peak, the township boasted a population of 5000 and five pubs. Nowadays the population is very small and all commercial mining has stopped. However, a few prospectors still strike it lucky, and the field is a popular spot with treasure hunters.

White Cliffs lies just over 100 kilometres to the north of Wilcannia, and about 295 kilometres north-east of Broken Hill. Access is over a rough

A magnificent crystal opal taken from the White Cliffs fields.

PHOTO DAVID BARNES, COURTESY NSW DEPARTMENT OF MINERAL RESOURCES. OPAL COURTESY SAPPHIRE AND OPAL CENTRE.

dirt road which can deteriorate in bad weather. There is a small village, but few facilities, and many of the residents live underground where they are insulated from the extreme temperatures. A walk-in mine has been constructed for the benefit of tourists and visitors.

In contrast to the black opal of Lightning Ridge, White Cliffs produces a predominantly light crystal opal. There is an absence of the nobbies associated with Lightning Ridge, most of the opal being found in thin vertical and horizontal veins. Since the colour is always in horizontal layers, opal taken from vertical cuts has the appearance of being banded with colour — an attractive and unique characteristic of White Cliffs opal.

Another interesting feature here is that the opal occurs in fossil shells or takes some other unique form rather than replacing wood, which is the more common formation. The entire geological scene at White Cliffs is interesting, since the sedimentary rocks have produced a wide range of fossils, indicating that this area, now more than 1600 kilometres from the coast, was once an estuary or shoreline.

Equally interesting to the geologist, but frustrating for the prospector, is the preponderance of the steel band layers of duricrust (mostly silcrete) close to the surface, which inhibit the excavation of opal dirt beneath. Sinking a shaft through a layer of silcrete could be likened to an attempt at penetrating one of Hitler's World War II bomb-proof submarine pens with an ice pick. The hardness of this layer of rock is indicated by the flat-topped mesas in the area around White Cliffs. Capped by silcrete, these protuberances, mostly less than 20 metres high, have remained protected while the surrounding sandstones and claystones have weathered and eroded away.

The main workings in the White Cliffs opal field are indicated on the map. There are a few smaller places nearby.

Mesas of different sizes and shapes are a feature of the opal-bearing regions of both South Australia and the White Cliffs area of New South Wales. The hard tabletop cappings of silcrete or ferrocrete have resisted the forces that have eroded and lowered the surrounding countryside.
PHOTO ROBYN HILL.

Gemville
Once called the Bunker field, this was one of the early fields to be discovered; it has produced good results over the years and is situated about 19 kilometres south-west of White Cliffs. One of its main advantages is that the silcrete capping is less extensive than at most other fields and therefore sinking a shaft is somewhat less traumatic. Most of the opal taken from this field has been from a depth of between 6 and 9 metres.

Barclay's Bunker
Erosion along Bunker Creek has exposed sediments with potential for producing opal. Shafts have been sunk in the area, which lies some 13 kilometres south-west of White Cliffs.

Walsh's Knob
A few shafts have been sunk, but this is a field badly inhibited by silcrete capping. It is 11 kilometres south of White Cliffs.

Purnanga
An old field about 48 kilometres north-east of White Cliffs. Although about 50 shafts have been sunk, the rock is hard and mining restricted.

SOUTH AUSTRALIA

Some 80 per cent of the world's opal comes from South Australia. Small wonder, then, that the fields to the north-west of Adelaide are popular with amateur fossickers. Andamooka, Coober Pedy and Mintabie lie close to the Stuart Highway, which runs north from Adelaide to Alice Springs and Darwin, making access generally easy, although roads in the opal fields adjacent to these centres can become difficult after heavy rain. For those who prefer to do it the easy way, Coober Pedy can be reached by regular air and coach services.

The precious opal found in the South Australian fields occurs in rocks that were broken down by weathering in the Tertiary period, somewhere between 1.8 and 70 million years ago. The weathering caused the rock to partially disintegrate, providing paths for water to percolate into faults and fractures. Acids in the water dissolved out fossilised shells and soluble minerals, creating cavities in the rock. Later, water containing silica seeped down through the faults and joints, depositing opal gel in the cavities. Many of the opal finds in the Coober Pedy area are in fossil shapes.

As in other states, there are two ways of obtaining opal in the South Australian fields — noodling and mining. Only the most dedicated prospectors take up a mining lease and sink a shaft in an attempt to find the rich seams of opal that lie beneath the surface. This involves hard work, a great deal of expertise and almost full-time dedication.

Many of the community who work the mines professionally live in the caverns they have carved from the rock — using the insulating effect of the earth to keep their dugout houses warm in winter and cool in summer. As in most Australian opal fields, extreme temperatures are experienced, particularly in summer.

A major opal field at Coober Pedy.
PHOTO MARIE AND BARRY AYERS.

COOBER PEDY

The first Australian opal was discovered as early as 1849, at Angaston, near Adelaide. Despite this, South Australia was one of the last states to begin commercial mining, the Coober Pedy deposits being discovered in 1915, some time after opal mining had begun in New South Wales and Queensland. However, because of the fine quality and prolific deposits of its opal, Coober Pedy soon became Australia's largest producer. Further development of the field eventually made it the world's greatest producer of the precious gem, a position it held until recently, when the output from nearby Mintabie overtook that of Coober Pedy.

The township of Coober Pedy sits astride the Stuart Highway, some 750 kilometres north of Adelaide, in the Stuart Ranges. The road is sealed all the way and the town has full facilities including shops, hotels and camping areas. The name derives from the Aboriginal *kupa piti*, meaning white man's burrow, a term used by the local Aborigines to describe the diggings of the early miners.

The opal field is extensive, stretching some 40 kilometres to the north, 10 kilometres to the south and 15 kilometres to the west of the township. Access to the different areas is mostly by rough dirt track.

ANDAMOOKA

Of the major South Australian fields, Andamooka, although closer to Adelaide, is more difficult to reach than Coober Pedy. From the Stuart Highway turn-off, about 400 kilometres north of Adelaide, access is over some 110 kilometres of rough dirt road, which becomes impassable to two-wheel-drive vehicles after rain. However, since it is a popular fossicking area for treasure seekers, a regular coach service operates from Adelaide directly to the small township of Andamooka. Here there are full facilities, including two motels and a camping ground, and tourists are catered for in the opal fields with specified fossicking and mining areas.

The opal field is gathered fairly tightly around the township, the main deposits lying within a radius of about 5 kilometres, although a few

deposits to the south-east extend almost 10 kilometres from the central area. The opal lies between two layers of claystone, the upper known as kopi and the lower as mud. It varies in depth, and mining is inhibited in some areas by deposits of silcrete.

MINTABIE

Currently the largest producer of precious opal in the world, Mintabie is very much a Johnny-come-lately to the opal scene. Although opal was discovered in the area in the 1920s, production was small and eventually the area was abandoned by all but small-time miners. In 1976 a new venture was begun, using heavy earth-moving equipment, and Mintabie came alive again. Big finds were made and renewed interest soon lifted the population from a handful of recluses to almost 500 enthusiastic miners.

The opal lies in a deep layer of sandstone from which it can be recovered by conventional underground mining. Heavy overlays of duricrust (silcrete) inhibit shaft sinking and commercial methods in which the surface duricrust is blasted and the layers beneath bulldozed out, have proved more successful. Noodling in the wake of the bulldozers can be a profitable pastime.

Although a licence is not required for fossicking, it is necessary to obtain permission to visit Mintabie as this is Aboriginal land. A permit can be obtained from The Administrator, Anangu Pitjantjatjaraku, 37 Bath Street, Alice Springs 5750.

QUEENSLAND

Queensland has the most extensive opal fields in Australia. They are not necessarily the richest, nor is the quality of the opal the highest, but the scattered fields cover a great area of the state and because of this, the exact extent of the deposits is not known. So Queensland compensates for its lack of concentration and quality of opal deposits by having a more exciting potential for future discoveries than any of the other opal-producing states.

Queensland's known opal fields lie within a belt of sedimentary rocks known as the Winton Formation, which extends from the New South Wales border in the vicinity of Hungerford, north-westwards over 900 kilometres of mostly arid country to the Kynuna region. For the most part it lies to the west of a line drawn between Cunnamulla and Winton and is about 130 kilometres wide, although this varies in parts.

The main opal mining centres are in Yowah, Toompine, Quilpie, Jundah, Opalton, Mayneside and Kynuna. The region encompassed by the opal belt is one of the hottest in the state, and summer temperatures frequently reach 40–42°C with extremes to more than 50°C in some places. Mesas are common, the level, hard capping of silcrete standing around 60 metres above the surrounding flat plains. As in all dry areas, care must be taken when prospecting, and only experienced fossickers, with suitable equipment and vehicles, should venture far from the centres of civilisation.

The opal is mostly recovered by sinking shafts, or in areas where silcrete inhibits vertical shafts, by driving angled or horizontal shafts

A busy scene at the pitheads of the Yowah opal fields.
PHOTO DR B. BRINSMEAD.

into the side of hills. For this reason, plus the harsh environment, prospecting requires a great deal of hard work in very difficult conditions. The Queensland fields attract only the most intrepid amateurs, and then only in winter, when climatic conditions are favourable.

Precious opal occurs beneath the silcrete in the softer layers of rock known as Grey Billy by old-timers. Sometimes it forms in boulders of hard ironstone concretions, when it is known as boulder opal. When found in a network of fine veins in the concretion, the whole mass of opal and ironstone is known as matrix opal. Other occurrences are at or near the junction of white sandstones or clays, and, in the Yowah area, as unique formations in which the opal is formed as a kernel in the centre of a spherical-shaped ironstone nodule. These are known as Yowah Nuts. There are other less common forms of Queensland opal, known as pipe opal, wood opal and sandstone opal.

Since the opal fields of Queensland are widely scattered across a vast area, it is not possible to mention them all here. Only the major fields with reasonable access for amateur fossickers are described.

One of the quite unique Yowah Nuts, split open to reveal its valuable kernel.
OPAL COURTESY DR B. BRINSMEAD.

YOWAH

This is the most popular opal field in the Queensland belt for amateur fossickers and tourists. In the winter months this field has a sizeable population and offers limited facilities in the way of shops, petrol and bore water. Access is along the Bulloo Developmental Road from Eulo, which is some 70 kilometres west of Cunnamulla along the same road. The easiest road into the field from the main road turn-off, 19 kilometres west of Eulo, winds north-westwards along Yowah Creek for about 40 kilometres before veering left and turning back to Yowah, another 30 kilometres along the track. Shorter access roads are available but are very rough, and in wet weather often impassable.

DUCK CREEK

Reached by entering the road that turns off the Bulloo Developmental Road, as described for Yowah, but continuing on and not taking the left turning to that field. Duck Creek, and other minor fields, are about 50 kilometres farther up the Yowah Creek. The roads here are rough and only accessible to four-wheel-drive vehicles and there are no facilities.

EROMANGA

In the high ground to the west and north-west of Eromanga are numerous mining areas, with some prolific mines, where amateur opal hunters have a reasonable chance of finding the precious gem. Access is through Quilpie, westwards along the Diamantina Developmental Road for about 50 kilometres to the turn-off for Eromanga. A further 70 kilometres along, a passable road leads to the township of Eromanga, and access to the mining areas. There are no facilities at the mine sites, so this is the last town in which to stock up with supplies. The roads beyond Eromanga are rough and mostly accessible only to four-wheel-drive vehicles.

KYABRA

There is no township of Kyabra, but turn-offs from about 6 kilometres north of the deserted homestead of Quartpot lead to the opal fields in the Canaway Range area, to the west of Kyabra. There are no facilities anywhere near these fields and only rough roads provide access. All supplies should be bought at Eromanga, about 80 kilometres south.

OPALTON

Coach tours take tourists out to the Opalton fields, so access is easier here than in many opal areas. There are a number of access roads, most of them rough and requiring four-wheel-drive vehicles even in good weather. There are access roads leading off the Winton–Jundah road at a number of points, but the most direct approach is from the Winton end where there are a couple of turn-offs just south of the town. These join a minor road which provides reasonable access to the mining areas, about 100 kilometres from the turn-off. There are no facilities in the area so everything required must be brought in from either Winton or Jundah.

A LITANY OF SHIPWRECKS

It is important to note that some of the shipwrecks that lie scattered around Australia's coastline are declared historic sites and must not be disturbed, damaged or removed. Every wreck is a part of Australia's heritage and must be protected and preserved as tenaciously as if it were a personal possession. This is particularly the case in the waters off the west coast where some of the wrecks are steeped in the history of Australia before the arrival of the white man. The thrill lies in the searching and the finding, not in the possession, and neither the wreck nor its environment must be disturbed, or a part of Australian history will be lost.

It is also important for amateur treasure hunters to understand the dangers associated with diving. While many wrecks lie in shallow water and can be explored with reasonable safety using only snorkelling gear, those in deeper water require the use of SCUBA equipment. Only divers thoroughly trained and experienced with this equipment should attempt to dive on wrecks in any depth of water, since lack of such training and experience can lead to fatal accidents. In most parts of Australia SCUBA gear can only be hired by persons holding a qualified diver's

A ship's telegraph.
PHOTO SCOTT CAMERON, COURTESY BORONIA ART GALLERY, MOSMAN.

Opposite: Many wrecks along the Western Australian coast lie in relatively shallow water, where they can be seen (but not touched) by amateur divers. Examination of wrecks such as the *Tryal* (1622) provide the ultimate in underwater excitement.
PHOTO PATRICK BAKER, COURTESY WESTERN AUSTRALIAN MUSEUM.

licence. And never go diving alone.

Excellent museums, such as the Western Australian Museum, are well equipped to protect and preserve historical wrecks and ensure that their remains are held in trust for all Australians. New discoveries should be reported immediately to such authorities. Ensuring that the maritime history of this nation is not lost, damaged or destroyed is a matter of conscience for every treasure seeker.

There are two major reasons why Western Australia is the favourite hunting ground for maritime treasure. Firstly, the coast lies, for the most part, behind an extensive string of reefs which run with few breaks from around Bunbury, north to North West Cape. Breaking the heavy surge of the Indian Ocean, these reefs create calm, clear water inshore — ideal conditions in which to search for lost treasures and sunken ships. By comparison, the much deeper, more dangerous southern and eastern coastlines of the continent limit the search for wrecks mostly to within a few hundred metres of the shore, and then only to experienced divers.

Secondly, many of the ships that plied along this coastline date back to the sixteenth century and were associated with the trade between Europe and the East. Many of the wrecks contain items other than the traditional treasures of coin and bullion, and the artefacts and relics from such ships have a value as part of Australia's maritime history. While it is always pleasant to find treasure that is financially rewarding, it is equally exhilarating and exciting to know that the wreck belongs to Australia, and that the discovery will add a little more to the slowly developing pre-history tapestry of this great country.

Not all wrecks occurred before settlement, of course, although undoubtedly those that did are the most interesting. But with the arrival of settlers at isolated spots around the Australian coastline, trade began, and because the land was so inhospitable and the distances so vast, trade by sea was the only viable method. The coastal waters, however, proved as inhospitable as the land itself, making every endeavour to disrupt this trade, and many fine ships found their last resting place in the turbulent white surge of ocean.

The sea is a possessive mistress and often it is in the last hours of a voyage, when thanks are being offered for a safe passage and celebrations begun, that she strikes with venomous cruelty. Thousands of men, women and children have reached the shores of this country after enduring the agonies and privations of a long voyage, only to have their first glimpse of their new home cut short by the waves that close over their heads. All shipwrecks are tragic but never more so than when lives are lost. The west coast, the south coast, Bass Strait and the east coast are the principal graveyards of the ships and their people that came so far but will never return.

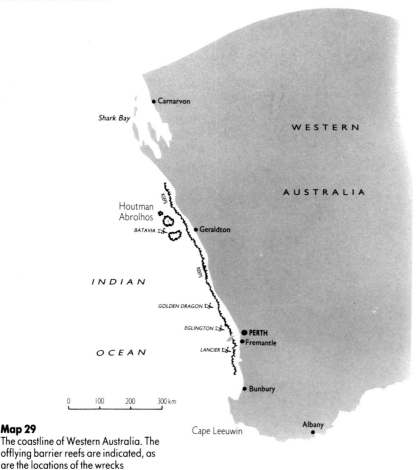

Map 29
The coastline of Western Australia. The offlying barrier reefs are indicated, as are the locations of the wrecks described in this chapter.

A NATIONAL TREASURE

In the days when, as the saying goes, "ships were wood and men were steel", the barque *Clan Macleod* was one of Australia's maritime shopping baskets. Just under 55 metres (180 feet) long, she was built of iron in England in 1873, and was to be one of the last of the solid jack-of-all-trades windjammers that did so much to develop Australia's trade with the world.

After 27 years of globe trotting, she was "retired" to the southern seas and renamed *James Craig*, operating mostly across the Tasman Sea, trading Australia's coal for a variety of goods needed to build the colonies into a robust nation. After a further 25 years, including four years of war service, her useful life was deemed to be over and she was taken to an isolated bay near the southern tip of Tasmania, and scuttled. The only thanks for her years of faithful service were a cold and lonely grave in a bleak backwater.

For forty years, the *James Craig* lay in the sheltered but cold waters of Recherche Bay. Her bow was above water, but her stern lay in about five metres, and the sea surged constantly through large rusted holes in her hull. Many of her 100-year-old plates were buckled and torn, but she had made herself a comfortable bed and was lying evenly on the bottom, with no undue strain on her structure. Fishermen used her to tie up their boats when sheltering from the ragings of the sea. Wildlife used her as a small and unexpected island in the middle of the wide expanse of the bay.

Then, one cold March day in 1972, she awoke from her long sleep to the tapping of surveyors' hammers, and the tickle of divers scraping at her thick coat of barnacles. A small, dedicated band of ship lovers from Sydney had come to lift the *James Craig* from her cold grave and breathe life back into her rusted bones. As one of the last authentic windjammers still recoverable from the shores of the continent, she was an important part of Australia's past — a national treasure.

The recovery was not easy. Without government assistance, the team had to rely on corporate generosity and the begging bowl to fund the highly involved repair work. Volunteers from every walk of life and from every state in Australia gave their time and their services to help bring the ship back to life. It was a race against time, for she was deteriorating badly. But gradually, centimetre by painful centimetre, her hull was lifted and pushed towards the beach. Then one day she was there, high and dry, her magnificent curved clipper bow towering over the sandy foreshore and her hull accessible for the final repair work that would enable her to float again.

On 26 May 1973, after months of repair work, the fine old ship poked her bow once more out into the ocean. Obedient to every move of the fussing tug that shepherded her through the shallows of the bay, she lifted to the ocean swell as she swung onto a northerly course, headed for Hobart for more repairs and, ultimately, her home port of Sydney. There was still a long way to go, for money was needed to make her seaworthy for the long trip north, and money of that calibre was beyond the scope of the dedicated band and their begging bowl.

Once again it was left to corporate generosity to save the ship. When the apathy of both Federal and New South Wales governments seemed certain to scuttle the brave venture, companies from across the nation, and notably from Tasmania, moved in. The enterprise had caught the imagination of the nation, and now money poured in to ensure that the *James Craig* would not only stay afloat, but would be restored to her full glory as the focal point of the Sydney Maritime Museum.

On 18 January 1981 Sydney Harbour was alive with a million small craft. In scenes usually reserved for royalty, or the start of the Sydney-to-Hobart yacht race, Sydney turned out in force to welcome its latest VIP. The stately bow of the *James Craig* dipped and bowed proudly as she entered Sydney Heads for the first time in more than half a century. Splattered by the wash of thousands of small craft and showered by the curving fountains of the fire-boat sprays, the fine old ship moved easily upstream under a bridge that had not been there when last she left her home port.

The *James Craig* was home, and one of Australia's national treasures had been saved.

EGLINTON

Life was exciting aboard the barque *Eglinton*, at sea in the Indian Ocean that night of 2 September 1852. After the long outward passage from Britain, the ship was at last closing the coast of Western Australia. The next day, if the wind held, they would see the shoreline of the new land, their new home, climb slowly up out of the sea. If their luck held, they might drop anchor in Fremantle Roads before dusk and, with even more luck, they might sleep ashore. A surge of excitement ran through the ship, a feeling the sailors knew well from many past landfalls. The twenty-nine passengers were in high spirits. Apart from the excitement of the approaching arrival, it was the birthday of one of their young ladies. Both events called for a celebration, so the steward and the cook decorated the ship's cuddy and produced a meal that was particularly sumptuous in view of the almost exhausted stores. The party got under way at dinner and carried on well into the night, both passengers and officers giving vent to their relief at the end of the long voyage. To add to their exhilaration, a fresh westerly wind was blowing, sending the ship creaming through the water, ensuring that she would arrive on time and that the celebrations would not be premature.

Captain Bennett, the ship's master, had set every stitch of canvas and it was drawing to its fullest extent. On deck, the helmsman moved the wheel easily and the ship responded quickly to the fast flow of water across her rudder. Aloft, the lookout peered ahead through the darkness, then relaxed and waited for the change of watch which was due any minute. Below, near the bulwark, he could see his replacement already reaching up to begin the climb up the ratlines to the crosstrees, so he stood, stretched himself and peered forward for the last time.

Something white flickered through the darkness. The lookout rubbed his eyes and peered again. It was not a light, but there was something faintly white out there. Then there was another, and another. He screamed the warning:

"Breakers ahead! Breakers ahead!"

It was too late even for the helmsman to put the wheel over. With a heavy thud, followed by a sickening crunch, the ship lurched to a standstill, throwing the crew out of their bunks and the celebrating passengers into a heap. Another lurch and ominous scraping sounds followed, while the crew scrambled aloft to reduce sail. Then another lurch and, miraculously, quiet. While everyone aboard held their breath, the ship shook herself and settled onto an even keel. She had sailed right over a reef.

The chain of reefs the fringe the coast of Western Australia and stretch almost its entire length from Cape Naturaliste to North West Cape, reach out sometimes as much as 20 kilometres from the shoreline. Inside, the water is calm and quiet, the surging swells of the Indian Ocean shattered into submission by the outer fringe of the reefs.

The *Eglinton* had, incredibly, crunched her way over the outer reef and was now lying in the calmer water inside, about 5 kilometres from the shore. In crossing the reef, however, the hull had been badly damaged and was making water. The rudder had been left behind, torn off by the impact. The crew reduced sail as quickly as possible, since the ship was still making way towards the shore, and without a rudder her movements could not be controlled. Before all the sail had been taken in,

Careful excavation with special equipment ensures that no part of the wreck or her treasure will be lost or damaged.
PHOTO PATRICK BAKER, COURTESY WESTERN AUSTRALIAN MUSEUM.

however, she struck another reef, less than 2 kilometres from the shore, and stuck fast. The captain ordered a gun to be fired. It was heard in Perth, a few kilometres to the south, but no help arrived, for no one dreamt it could be a distress signal.

Captain Bennett set about abandoning the ship without haste or panic. He had two problems — getting the passengers and crew off safely, and attempting to salvage some of the cargo. The first was a normal seamanship exercise, and although two people were drowned, this was hardly the fault of the ship's personnel. The bosun got drunk and threw himself in the water and one of the passengers fell out of the ship's boat when landing and was drowned in the surf.

The cargo presented a different problem, however, for in the after hold, among various items of normal cargo, were chests containing 65,000 sovereigns intended to boost the coffers of the State Treasury. Even at this early stage, some of the crew had shown signs of becoming unruly, for after the passengers had been put ashore, three of them raided the passengers' cabins and plundered their belongings. Captain Bennett decided the best move was to get every person off the ship and wait until help arrived.

It was not until 6 September, three days after the wreck, that assistance in the form of the Fremantle police boat and a chartered schooner, the *William Pope*, arrived off the wreck of the *Eglinton*. By this time she had settled by the stern with her bow in the air and the water level with her after deck. A diver was put aboard from the *William Pope*, which had been fitted with salvage gear, and the work of recovering the cargo began. Inspector Clifton of the water police and Captain Wray of the Royal Engineers directed the operations.

Although at first the diver had difficulty getting to the money chests because the cargo had been thrown around when the ship grounded on the reef, eventually they were located. Just as the sun was setting, two chests were brought up from the flooded hold and lowered into the police boat.

The *Eglinton* broke up soon afterwards, and the beaches along the shoreline were strewn with debris. Just how much was saved and what was left in the ship when she broke up is not known. Because she was wrecked close to the shore where the water is relatively shallow and diving is easy, searching for any scattered remains of this ship will provide a good incentive for beginners and novices to learn the skills of diving on a wreck in ideal conditions. It is important to remember, however, that the care and ownership of all wreck sites is vested in the Western Australian Museum and nothing may be touched, disturbed or removed. The Museum should be contacted immediately if any new discoveries are made.

Treasure from the 1811 wreck of the *Rapid*, including Spanish dollars. All such treasure is a part of Australia's heritage and belongs to the nation. Individual stimulation and reward for this kind of treasure hunting comes in the finding, not the keeping.

PHOTO PATRICK BAKER, COURTESY WESTERN AUSTRALIAN MUSEUM.

LANCIER

On 25 August 1839, the 285 tonne barque *Lancier*, under the command of Captain Durocher, sailed out of Port Louis, Mauritius, bound for Hobart via the western seaport of Fremantle. In her holds was a mixed cargo of 1782 bags and 126 casks of sugar, rum, soap and a variety of "sundry" items. There was no mention on the manifest of a chest of silver coins believed to be worth £7,000.

The voyage was uneventful and the barque arrived safely off the island of Rottnest in the early morning of 28 September. In accordance with harbour regulations, she signalled for a pilot to take her into the port of Fremantle, then hove to. After waiting for some hours with no acknowledgement of his signal, and no sign of the pilot boat putting out, Captain Durocher became impatient and decided to take the ship in himself.

A man of wide navigational experience in many parts of the world, he was more than capable of negotiating the entrance to the harbour, and, to the best of his knowledge, his charts were accurate and up to date. A light breeze was blowing, so without further delay, Durocher turned the *Lancier*'s bow towards the shore and began his approach to the entrance.

Posting a man in the chains to sound the water and a lookout to watch for hazards up ahead, Captain Durocher decided to enter through the channel to the south of Rottnest Island. Today the channel is clearly marked with buoys and beacons, but the *Lancier* had only her charts to indicate problems ahead. These showed two rocks, one with just over a metre of water on it and the other with about 2 metres of water. Between the rocks there was deep water on all sides. Sailing steadily past the first rock, the ship appeared to be in no danger at all. The lookout reported all clear ahead and the leadsman reported almost 10 metres of water under the keel.

But the charted position of the second reef was incorrect, and with a violent crash that threw everyone aboard off their feet, the *Lancier* impaled her bow on the sharp rocks and immediately began to fill with water. Knowing that the ship was doomed, Captain Durocher ordered more sail set in an effort to drive her farther on to the reef and thus provide a survival platform for those on board. But she was locked fast, and as the water in the holds rose, there was danger that she would either break up or sink stern first into deep water.

The awkward angle at which the ship was wedged made launching the boats almost impossible. The crew did manage to get one into the water, and passengers and crew began to abandon ship. But the boat was not big enough for them all and soon became dangerously overloaded. It was decided to make in to the beach and call for assistance, leaving four of the crew clinging to the wreck.

At this point the mystery of the chest of coins comes to light. The whaleboat that put out from the shore to rescue the remaining survivors was under the command of a part-time pilot, Captain Dempster. On the way out, the rescue boat passed the heavily laden ship's boat making in to the shore. Captain Durocher hailed Captain Dempster and asked to be taken back to the wreck, but Dempster refused and continued to the rescue of the stranded seamen.

In his log, Dempster noted that when his whaleboat reached the wreck, the four seamen had on deck a chest full of silver specie. Was this why Captain Durocher was so keen to return to the wreck? Was he attempting to salvage the most valuable part of his ship's cargo? If so, why was it not listed on the cargo manifest? Or was it some secret shipment of coin Durocher was attempting to smuggle into the country?

The answer will never be known, but Captain Dempster is quite definite in his description of the chest and its contents. Since there were only four sailors to rescue, he could easily carry the chest, so the crews set about loading it into the boat. Exactly what went wrong is not recorded, but in transferring the chest from the *Lancier* to the whaleboat, it slipped from the men's hands and plunged to the bottom. That night a

Superb examples of "Beardman" jugs, some of which were recovered from the *Golden Dragon*, also known as the *Gulden Draak*. After more than 300 years on the sea bottom, only treatment by skilled technicians prevented them from suffering damage or even total loss.

PHOTO PATRICK BAKER, COURTESY WESTERN AUSTRALIAN MUSEUM.

rising sea caused the ship to fill completely and she slipped backwards off the rocks and sank in deep water.

The loss of the treasure was reported in the Perth newspaper at the time, yet no official explanation has been offered as to why it was not recorded on the ship's manifest. Apart from theories on Captain Durocher's attempt to smuggle the coin into the country (which does not make sense in the light of the currency situation at the time), there seems to be no logical reason why it would have been excluded. Was there really a chest of specie? If so, why didn't Captain Dempster pursue it? With his knowledge of the local waters, finding and raising the chest should have been a relatively easy task. One story has it that the pages of his log containing information about the whereabouts of the treasure had been deliberately torn out. If this were so, then perhaps the treasure has already been recovered, although no record of this has come to light, either officially or along the grapevine of the Fremantle waterfront.

The remains of the ship have been located, together with another, the *Zedora*, in the same area. Relics from the *Lancier* have been raised and are housed in the Western Australian Museum at Fremantle. But of the treasure chest of silver coins there has been no sign. If there was such a chest, it may still be awaiting discovery somewhere beneath the clear water off Rottnest Island.

Recovering relics from the wreck of the *Golden Dragon* (1656).
PHOTO BRIAN RICHARDS, COURTESY WESTERN AUSTRALIAN MUSEUM.

GOLDEN DRAGON

One of the most exciting finds along the west coast has been that of the *Vergulde Draeck* or *Gulden Draak*. Better known in Australia as the *Golden Dragon*, she was one of the biggest of the Dutch East India ships at the time, with a displacement of 600 tonnes and an armament of 30 cannon. On her second voyage to Batavia she was believed to be carrying one of the richest cargoes ever lost on the Western Australian coast. It included at least eight chests of coins, as well as the personal valuables of her 187 passengers and crew.

The *Golden Dragon* left Texel, Holland, on 4 October 1655. She was on her second voyage to Batavia under the command of Pieter Albertsz, an experienced skipper who was very familiar with the passage to the east and who was a keen advocate of the Southern Ocean route. On 28 April 1656, the ship closed with the coast of Western Australia at a point some 110 kilometres north of the present site of Perth.

Until then it had been a trouble-free voyage, but the treacherous reefs of the west coast of Australia still had to be negotiated. Just after midnight, the *Golden Dragon* struck one of these reefs with enormous force, smashing in her forefoot and filling quickly with water. There was immediate chaos. Passengers and crew panicked in their attempts to save their belongings and climb into the boats before the ship broke up. Some remained aboard in the hope that she might stay afloat. But a strong wind was blowing, causing great breakers to surge around the stricken hull. In a very short time, the magnificent East Indiaman was reduced to a broken hulk. She slipped stern first beneath the waves, leaving wreckage and drowning survivors strewn across the 8 kilometres of water between the reefs and the shore.

Of the 187 souls aboard, 117 perished within those first disastrous hours. Seventy reached the shore, only to be confronted with an equally desperate situation. There was little likelihood of rescue, since passing ships kept well off this coast in order to avoid the very reefs on which *Golden Dragon* had foundered. The land on which they stood appeared arid and barren, with little sign of water. Their sole means of obtaining help was the small ship's boat which had survived the chaos of the wreck.

Captain Albertsz decided to remain with his survivors and instructed one of his officers to sail the ship's boat northwards to find help. The boat and its crew of seven reached Batavia after a remarkable feat of seamanship and raised the alarm. Two ships, the *Witte Valk* and the *Goede Hoop*, were immediately dispatched southwards to search for the survivors and salvage any of the *Golden Dragon*'s treasure that might be recoverable. But rough seas and the risks involved in penetrating the almost continuous line of offshore reefs made the task a difficult one. When the ships did get close to the shore, they found wreckage from the ill-fated East Indiaman strewn all along the beach, but there was no sign of any survivors. The wreck itself had completely disappeared, so there was no hope of salvaging the cargo. The *Goede Hoop* put a search party ashore to scour the hinterland, but they did not return, and despite numerous attempts to find them no trace could be found of either the survivors or the search party. Another eleven names had been added to the already tragic list of victims. Further searches followed, although little hope of finding the castaways or the treasure now remained, and the rescue ships finally gave up and returned to Batavia.

But the tragedy had not yet run its full course. Less than two years later, another expedition left Batavia for the coast of Western Australia. Two ships, the *Waeckende Boey* and the *Emeloot*, sailed southwards and began a search of the coastal region where the *Goede Hoop* had found wreckage. The shore parties from this expedition also found wreckage scattered along the shore from the present site of Guilderton, north to the present location of Lancelin. There was no question that it was from the *Golden Dragon*, but there was no sign of survivors.

One night, while a shore party scoured the beach near Lancelin, the wind got up and the *Waeckende Boey* had to weigh anchor and sail away from the shore to avoid being driven onto the reefs. When she returned in daylight, there was no sign of her boat or the fourteen men of her search party. After landing further search parties and sailing up and down the coast for a few days, the skipper decided the boat had capsized in the blow and the men were lost. He turned his ship's head out to sea and sailed back to Batavia.

One hundred and eighty-five days later, four survivors of the boat's crew staggered into Batavia. Having watched their ship leave without them, they had sailed more than 3000 kilometres in their tiny boat, losing ten of their comrades either as a result of drinking seawater or at the hands of hostile natives on islands where they had landed. The seaman in charge of that epic voyage was Abraham Leeman, and his name is perpetuated to this day in a small fishing village on the coast where he and his crew were marooned in 1658.

The Dutch gave up the search that had cost so many lives, and the wreck of the *Golden Dragon* and the treasure she carried to the bottom of the sea passed into history. The story of the "golden ship" effectively became a legend as almost three centuries passed. Then, in 1931, coins dating back to the mid-seventeenth century were found on the beach

Remains of the *Batavia* on the Houtman Abrolhos reefs. A diver from the Western Australian Museum examines the hull timbers which, after more than 300 years of submersion, still clearly show the ship's structure. Careful and responsible recovery will enable this ship to be seen by tens of thousands of Australians in the Western Australian Maritime Museum in Fremantle.
PHOTO JEREMY GREEN, COURTESY WESTERN AUSTRALIAN MUSEUM.

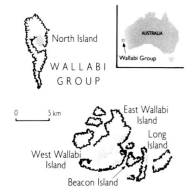

North Island

WALLABI
GROUP

AUSTRALIA

Wallabi Group

0   5 km

East Wallabi
Island

Long
Island

West Wallabi
Island

Beacon Island

near Guilderton. These proved to be Spanish and Dutch coins, presumably washed ashore from an old wreck. Interest in the numerous offshore wrecks was rekindled, and enthusiastic treasure seekers searched the reefs to the north of Perth, but without success.

Then, in 1963, a party of amateur divers located a wreck about 8 kilometres offshore, not far from Guilderton. The cargo and a number of artefacts in the hull seemed to indicate that this, indeed, might be the missing *Golden Dragon*. Following vandalism from irresponsible treasure seekers, the Western Australian Museum placed the wreck under a protective order until a full-scale expedition could be mounted. In 1972 the site was excavated by the Museum, and all doubts about the identity of the wreck were dispelled. The *Golden Dragon* had at last been found.

AUSTRALIA'S FIRST MUTINY

The most famous of the wrecked Dutch East Indiamen lies on the offshore coral reefs to the west of Geraldton, almost 500 kilometres north of Perth. These extensive reefs, and the islands scattered among them, are known as the Houtman Abrolhos after the Dutchman who discovered them. The ship is the *Batavia*, one of the finest ships in the Vereenigde Oost-Indishe Compagnie (V.O.C., or Dutch East India Company) fleet. Her remains were discovered in 1961, some 332 years after she had broken up and sunk after being stranded on the coral reef. Since then a wealth of treasure of both intrinsic and historic value has been raised from her hull by expeditions from the Western Australia Maritime Museum.

There is an interesting parallel between the story of the *Batavia* and that of another well-known ship, HMS *Bounty*. Both ships were sailing in southern oceans, albeit some 160 years apart, in similar latitudes. The *Bounty*, with Captain William Bligh in command, was to the east of Australia in 1789, and the *Batavia*, skippered by Captain Adriaan Jacobsz, was to the west, in 1629. Both captains were reputed to be tyrannical commanders and both were involved in a mutiny.

Both captains set out for assistance, each sailing a small boat across hundreds of kilometres of little-known ocean. Bligh headed westwards around the north-east tip of Australia to reach Timor, while Jacobsz sailed eastwards around the north-west tip of the continent to Batavia, in the same region. Both these voyages were to be hailed as feats of seamanship and endurance of the highest order.

In both cases, most of the mutineers were finally caught and hanged, and the ships on which the dramas were played out found their final resting places on the seabed of a remote tropical island. The remains of the *Batavia* lie among the coral outcrops of the Houtman Abrolhos in the Indian Ocean, and the charred ribs of the *Bounty* can still be seen beneath the waters of a quiet bay on Pitcairn Island in the Pacific Ocean.

The story of the *Bounty* has been told and retold innumerable times but with the exception of a couple of books, and possibly a few magazine articles, the story of the *Batavia* has received little attention. And yet in terms of horror, inhumanity and depravity, few tales of the sea can approach this true story of Australia's first mutiny.

It began when the *Batavia* sailed from Texel, Holland, in 1628, in company with two other ships. They formed part of an expedition of

More than three centuries after her tragic loss, the *Batavia* takes shape again in the Western Australian Maritime Museum, as hull timbers recovered from her watery grave are brought ashore. The stone archway was part of her ill-fated cargo, destined for the city of Batavia, Java.
PHOTO PATRICK BAKER, COURTESY WESTERN AUSTRALIAN MARITIME MUSEUM.

eleven ships bound for the eastern port of Batavia, after which the ship had been named. The expedition was under the command of François Pelsaert who sailed aboard the *Batavia* with her captain, Adriaan Jacobsz and 316 passengers and crew, including another officer of the V.O.C., Jeronimus Cornelius.

Trouble plagued the ships from the start. Bad weather struck before they reached the Cape of Good Hope and the convoy was scattered. This was not an uncommon occurrence in those days, and the *Batavia* sailed on alone, intending to meet up with the rest of the fleet at their destination. Once having rounded the Cape, she followed the standard route sailing eastwards across the southern Indian Ocean, before turning north to pick up the trade winds that would carry them up the coast of Western Australia, or the South Land, as it was known in those days.

Bad blood between Pelsaert and Jacobsz created an unpleasant atmosphere on board. Pelsaert was in command of the convoy and therefore Jacobsz was his subordinate. But Jacobsz had previously been senior in rank to Pelsaert, who had been instrumental in Jacobsz' demotion. From the outset the recalcitrant skipper had sworn revenge, and by the time the ship reached the coast of the mysterious South Land, the atmosphere on board was heavy with the threat of mutiny.

On the night of 4 June 1629, fate took a hand. A lookout reported white water up ahead. It was dismissed by Jacobsz as moonlight on the water and the *Batavia* sailed on under a full press of canvas. Shortly afterwards, with a terrifying crash, the ship ran hard aground on an uncharted coral reef. Pelsaert, who had been asleep in his cabin, raced on deck to find utter chaos, passengers and crew milling around hysterically as waves surged across the deck. He immediately set about calming the people and restoring order.

After the initial panic, it became obvious that the ship would not break up immediately, so Pelsaert ordered passengers and crew to abandon ship in an orderly fashion and make for a small island which was just discernible in the light of the approaching dawn. But there were insufficient boats and the transfer of everyone on board to the safety of the shore was a slow and dangerous task.

In the hope that she had grounded at low water, the crew began off-loading all the heavy equipment so the ship might float as the tide rose. In fact, she had struck at the top of the tide, and as the water fell away from her hull it was obvious that she was firmly held and would start to break up if the weather turned. Already the surge of the ocean was rending her timbers and she was soon beyond any hope of recovery. The aim now was not to save the ship but to save as many lives as possible. Already some had perished attempting to swim ashore.

All hands were put to work retrieving as much food and water as possible from the wreck and ferrying it to the small island. Daylight revealed other islands nearby. Since they lay near the established sea route to Batavia, their chances of survival began to look better. Pelsaert set about establishing a survival camp on the largest island, ferrying more supplies from the still visible wreck.

When the survivors took stock of their situation it was obvious there was sufficient food to last for some time, but lack of water posed a problem. An extensive search of the adjacent islands failed to discover any source of water and the situation began to look grim. It was decided that Pelsaert and Jacobsz, together with a few of the crew, would sail the ship's boat to Batavia for assistance. In what is now hailed as one of the finest feats of seamanship in maritime history, Jacobsz navigated the tiny craft northwards around the coast of Australia and through the Indonesian islands to Batavia, arriving safely exactly one month after the wreck of his ship.

While Pelsaert and Jacobsz were away, the survivors, who were now spread over three islands, had been left in the charge of Jeronimus Cornelius. But, unlike the others, Cornelius had no intention of sitting and waiting to be rescued. Among the cargo transferred from the wreck to the island were a number of chests of coins. Cornelius planned to organise a small gang of pirates from among the stranded crew, then seize any rescue ship that might appear and take off with the *Batavia*'s treasure, to embark on a life of piracy on the high seas. To ensure that there were no obstacles to his plan, Cornelius devised a horrendous scheme. With a small group of followers he set about systematically murdering all men among the survivors who would not join his group and eradicating any women he or his crew did not want for their pleasure. In this way he planned to have a loyal band organised and waiting to capture the rescue ship when Pelsaert returned.

The massacre began secretly. Several young men were bound and taken out to sea where they were pushed overboard and drowned. At the same time, a number of people living on one of the nearby islands were annihilated. But the remaining survivors became suspicious, so Cornelius abandoned all attempts at secrecy and ordered his men to embark on a rampage of rape and murder. When they had finished, 125 people were dead and only a small group on a distant island remained to be dealt with.

This group, learning what had happened when a survivor of the massacre swam across, immediately formed a defensive party under the leadership of a soldier, — Webbye Hayes. But they numbered only 45, including women and children, and would obviously be no match for Cornelius's armed and organised thugs.

But Hayes was a shrewd commander and, using his tiny force cleverly, he beat off two attacks, causing Cornelius to withdraw and lick his wounds. On the third attempt, Hayes placed his men strategically along the beach, keeping them hidden until Cornelius stepped out of his boat and onto the sand. At this point he was most vulnerable and Hayes' men moved in swiftly and decisively. In the ensuing struggle, Cornelius was captured and his landing party killed.

On his return from Batavia in a rescue ship, Pelsaert searched for some time before locating the survivors. His horror at discovering the turn of events can only be imagined. He immediately set up a court of inquiry and tried and then hanged the murderers, including Cornelius, thus carrying out the first known executions on Australian soil. Two of the men, who were considered less guilty than the others, escaped the gallows but were cast ashore somewhere on the mainland. They were never seen again.

Pelsaert was able to recover much of the valuable cargo from the wreck of the *Batavia*. Indeed, as far as is known, only one chest of coins was left behind, since it was securely jammed in the wreck and could not be moved. When the wreck was examined in 1961 the chest was not found, nor was it recovered when further expeditions removed many of the relics in 1971. However, a number of rix dollars were found spilled across a part of the hull, giving rise to the theory that the wooden cask had disintegrated, leaving the coins exposed.

The story of a successful treasure hunt

The major problem that confronted the many adventurers who, over the years, have searched for the *Batavia* was the exact location of the wreck. The Houtman Abrolhos Group consists of widely scattered coral reefs surmounted at points with small islands. They extend over an area of about 80 kilometres in a roughly north-north-westerly direction and are divided into three main groups — the Wallabi Group to the north, the Easter Group, and the Pelsart Group to the south. There is a channel some 12 kilometres wide between the two northern groups, and a similar channel, about 7.5 kilometres wide, between the Easter and Pelsart Groups.

Initially, it was assumed that the *Batavia* had grounded on the reefs of the southerly Pelsart Group, and this theory was supported by wreckage found in the area. Indeed, the south-west tip of Pelsart Island was named Wreck Point on the assumption that it was here that the ship had met her end. But the wreckage, and remains of human habitation on the island, were later proved to be from another ship, possibly the *Zeewyck*, which was wrecked in the area amost 100 years later.

The exact location of *Batavia*'s wreck was resolved by the dedicated research work of noted Australian historian and author Henrietta Drake-Brockman. It was she who first realised that the generally accepted location of the wreck near Pelsart Island was not correct and that the relics found on that island were from another ship. Using information from translations of Pelsaert's own journals, Mrs Drake-Brockman determined that the description of the southern islands did not match those given in the journals. Further investigation convinced her that the wreck had occurred in the northern, Wallabi Group, although no relics which could be linked to the *Batavia* had been found there.

A breakthrough came in 1960 when a fisherman discovered human remains on Beacon Island, one of the Wallabi Group, which Mrs Drake-Brockman had pinpointed as the possible site. This aroused considerable local interest and shortly afterwards more finds were made which confirmed that this was indeed the *Batavia*'s Graveyard, as the survivors had named it. Divers discovered the remains of the ship in about 4 metres of water and it appeared that although much of the hull had disintegrated, there was a huge treasure in relics and artefacts to be recovered from the seabed.

A number of private expeditions examined the wreck, including one in 1963 on which Mrs Drake-Brockman was able to enjoy the fruits of her years of dedicated research. But so important a discovery required careful handling and in 1971 an expedition was mounted by the Western Australian Museum to recover as much of the remains as possible and transfer it into safe care at the museum's Fremantle establishment. A 2.5 tonne bronze cannon, two 1 tonne iron cannons, a 1 tonne anchor and numerous artefacts and coins were among the important items recovered and now preserved in the museum. But probably the most exciting find was a collection of navigational instruments, including an astrolabe, one of only ten known to exist in the world today.

The *Batavia* was but one of a number of vessels wrecked on the reefs of Houtman Abrolhos. Lying directly in the path of the unsuspecting Dutch

East India merchant vessels heading for eastern ports, the reefs claimed many a fine ship in the days before Frederik de Houtman recorded their position in 1619. And even since 1622, when they first appeared on Dutch charts, numerous unwary navigators, like Adriaan Jacobsz, have fallen foul of them.

The number of wrecks on the Houtman Abrolhos make them a prime treasure-hunting ground. Even modern ships, with the latest navigational aids, come to grief on the reefs lying so far off the western shore of the continent. However, it is important to note that any new wrecks discovered on this, or any other site, must not be touched until the Western Australian Museum has been notified and has cleared the wreck as being historically insignificant.

DISASTER AT GREEN CAPE

Because of the discomfort of covering the long distances overland, most travelling in nineteenth-century Australia was done by sea, particularly between the capital cities, all of which are on the seaboard. Usually, weather and sea conditions are good off the Australian coastline, and a journey between capital cities was a relaxing holiday for the day or two involved. A far cry from the noisy, smoky, long and uncomfortable railway journey or the even longer and more uncomfortable trek by road. Rival shipping companies fitted out their ships luxuriously to entice paying passengers, and first class, or saloon, accommodation was comparable with the best capital city hotels.

Passengers enjoyed superb service during their voyage, and the coast, as long as the weather was pleasant, provided a magnificent panorama of passing scenery. Many ships sailed upriver well inland to rural centres to eliminate the need for passengers to make the overland journey to the coast. Towns as far in from the coast as Lismore and Grafton became major seaports with regular passenger services to Sydney and other destinations.

Competition between the shipping companies operating these services ensured not only excellent accommodation and service, but also fast, well-built ships that kept to strict timetables and were only late when heavy storms lashed the coastline. Even then, ships were slowed more often for the comfort of passengers than for safety, for these vessels were among the best in the world and even the storms moving up from the Antarctic made little impact on their progress.

One of the prominent shipping companies of the day was the Australian Steam Navigation Company. A pioneer in coastal shipping services, the ASNC had begun its first services in 1839 with three paddleships — the *Rose*, *Shamrock* and *Thistle*. As trade expanded, newer and faster ships were introduced to upgrade the passenger runs between Brisbane, Sydney, Melbourne and New Zealand. The company became renowned for its punctuality and excellent accommodation, importing from Europe the best vessels of the time to maintain its high standards.

But luck was not always with the Australian Steam Navigation Company, for despite its fine record, two of its best ships share a watery grave on the south coast of New South Wales. They carried with them beneath the waves 71 souls and tens of thousands of dollars worth of goods and belongings, none of which has ever been recovered.

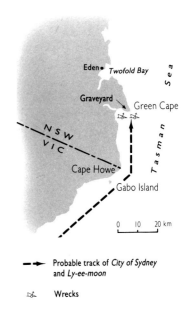

Map 31
The coastline in the vicinity of Green Cape, showing the probable tracks of the ill-fated ships.

The treacherous coast near Green Cape, which protrudes menacingly out into the shipping channels.

Green Cape was where these two fine liners sank to the bottom. A pointed promontory on the coastline just south of Twofold Bay, Green Cape creates a hazard for unwary mariners, jutting out into the shipping lanes that round the south-east corner of the continent. So dangerous is it that one of the major lighthouses on the New South Wales coast blinks a nightly warning from its tip to this day. Beneath it lies the ominously named Disaster Bay.

On the night of 5 November 1862, the fast ASNC steamer *City of Sydney* rounded Gabo Island and headed north for Green Cape. She was bound from Melbourne to Sydney with some 100 passengers aboard. Soon after midnight, fog closed in and the ship's master, Captain Moodie, altered course to seaward to allow more room off Green Cape. But his actions were too late and minutes later breakers were seen ahead. Before the helm could be put down the ship struck the rocks at full force and began to sink immediately.

Luck was with the survivors, for the night was calm and all passengers and crew were able to leave the sinking ship without even a minor injury, although one of the boats was almost smashed by the ship's mast when she rolled over and sank. The survivors were rescued by whale-boats from Eden, but their personal belongings, together with a valuable cargo, were lost forever beneath the waves.

Twenty-four years later, another fine vessel of the ASNC's fleet followed the shipping lanes around Gabo Island and headed north to Green Cape on the same course as the ill-fated *City of Sydney*. This was the crack steamer *Ly-ee-moon*, considered one of the finest and fastest

The rocks which tore the hearts out of two fine ships, and claimed 71 lives — Green Cape, seen from the top of the lighthouse. The lighthouse keepers' cottages were converted into a casualty station for the survivors. Many of the dead are buried in a tiny cemetery nearby.
PHOTO TOM BUDDEN.

passenger ships on the Australian coast. Built in England as a paddle steamer, she was capable of the remarkable speed of 17 knots, which made her the fastest vessel of her type in the world. The *Ly-ee-moon* had undergone a refit in Sydney and entered the Melbourne–Sydney trade in 1878 as the most up-to-date vessel of her era.

On this occasion she had left Melbourne on Saturday, 29 May, and was headed at full speed for Sydney where she was to berth early on Monday morning. Her large complement of passengers consisted mostly of holiday-makers going north for the coming winter, among them a number of children. On the evening of 30 May, she had rounded Gabo Island as the passengers sat down to dinner, and headed north towards Green Cape in a moderate to fresh wind, with rising seas. The master, Captain Webber, having negotiated the ship round the "corner" retired to rest, leaving the Third Officer in charge of the bridge.

Exactly what happened will never be known, for visibility was good and the light on Green Cape clearly visible. Yet at around 9 p.m. the ship smashed onto the rocks beneath the lighthouse and within three minutes broke in two. The after part of the hull slid backwards and sank; entombed inside it were 19 steerage passengers and the entire engine-room crew. The front section was battered by the seas pounding on the rocks and many of the survivors who had been clinging desperately to the wreck were washed away and drowned.

The lighthouse keepers raised the alarm and set about rescuing as many survivors as possible. Getting a line onto the wreck, they eventually pulled 13 of those remaining onto the rocks and safety. Despite working through the night, no further survivors could be rescued, even though cries for help were often heard, either from the wreck or from the water. The next morning 24 bodies were washed up on the beach. They are buried in a small cemetery on Green Cape marked by white stones.

The rest of the *Ly-ee-moon*'s hull disappeared beneath the waves to join the stern section and, not far away, the remains of the *City of Sydney*. No formal attempt at salvaging any parts of either ship has been made, although to this day divers occasionally come across a relic of that fateful double disaster which gave its name to the adjacent curve of beach — Disaster Bay.

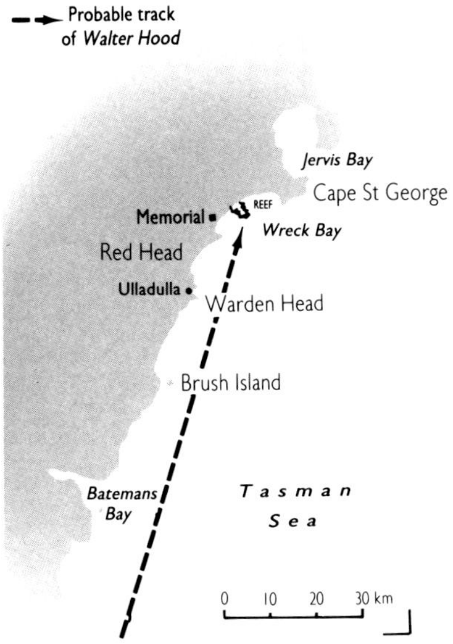

— Probable track
of *Walter Hood*

Jervis Bay
Cape St George
Memorial ■ REEF
Wreck Bay
Red Head
Ulladulla ●
Warden Head
Brush Island
Batemans
Bay
*T a s m a n
S e a*
0   10   20   30 km

Map 32
The *Walter Hood* had drifted so far inshore from the shipping track that she was doomed, even if she had missed the reefs which claimed her.

WALTER HOOD

The small, but well-sheltered port of Ulladulla lies about halfway between Port Jackson and Twofold Bay. Snugged up under Warden Head, it was a focal point for early settlers and farmers on the south coast, for until the Princes Highway was built, the shipping services into Ulladulla provided the only link between these isolated communities and the growing city of Sydney. Regular steamer services carried mail and supplies to the region, cut off from the rest of the colony by the Shoalhaven River to the north, the Clyde River to the south, and the impassable mountains inland.

In mid-April, 1870, the steamer services were cancelled. A fierce gale had battered the south coast, building up huge seas which crashed onto the headlands, sending spray flying like a heavy mist over the light-houses. The ships in the tiny harbour doubled their moorings on the solid stone pier, or ran out extra anchors to counter the surge that rolled around Warden Head. Fishermen battened their doors against the shrieking wind and huddled closer to the fire. It would be some days before they would be going out to sea again.

On the night of 23 April, to the south of Warden Head, the black silhouettes of a fine clipper's masts swept across the dark sky. Her rigging was stressed to breaking point, tatters of her topsails pointed raggedly ahead and her hull was racked and twisted as she corkscrewed violently in the huge swells.

But far from being concerned by the anger of the storm, the crew of the 950-tonne clipper *Walter Hood* were jubilant. They had ridden out many a worse blow than this, and since it was a "gung" southerly, it was blowing them at a speed of 12 knots towards their destination. Tomorrow they would be in Old Sydney Town after a fast passage of ninety days out from London. There would be wine, women, song and celebration and a bonus in their pockets for a fast delivery of their cargo.

The ship's master, Captain Latto, was happiest of all, standing on the darkened poop, feet braced against the wild plunging of the ship, his seaman's eye roaming aloft for the sign of something amiss. He was a typical clipper captain; a hard skipper, feared but grudgingly admired by his crew, for he drove his ship and crew to the limit of their endurance, but in doing so won fine honours and good bonuses from grateful shipowners and merchants.

Off Ulladulla, Captain Latto detected a distinct drop in the force of the wind. The New South Wales coast was renowned for its infuriating head winds, which often made the last part of a voyage the longest and most trying. Latto had been delighted to pick up the roaring southerly as the *Walter Hood* had sailed out of Bass Strait. It would be aggravating to lose it now, so close to their destination.

George Robinson, a settler at Ulladulla, also noticed the drop in the wind and sighed with relief. The storm had interfered with his work. Although the roads were muddy, things would start to dry out tomorrow, and he should be able to get his produce down to the harbour ready for the first ship to put into the port. He looked out of his windows and saw fog settling down in the gullies and out to sea. A sure sign that the storm was over.

But if the storm was over, the fury of the sea was not spent. The crewmen of the *Walter Hood* cursed as they clung to the gyrating masts and yards, 46 metres above the deck, attempting to cut free blown sails and set new canvas. On deck, Captain Latto paced back and forth with furious frustration. Unless they made a lot of progress this night, they would not reach Sydney tomorrow after all. He vented his frustration on the men aloft, bellowing commands and curses into the wildly plunging rigging. Despite the fog that was closing round them, he was not concerned for the ship's safety, for he knew he was well offshore. His sole thought was to get her sleek hull moving again.

An hour or so later, as the massive spread of canvas aloft slatted madly with the violent motion of the ship and the lack of wind to hold them full, the ship was lifted on a huge wave. The crew knew instinctively that this was not a natural wave. It wasn't. The *Walter Hood* was in shallow water. The next big wave lifted her high and crashed her down with a back-breaking jar on a rocky outlying reef.

There was no time to take evasive action. The next giant wave poured aboard, slamming Captain Latto against the bulwarks and bringing down

Top: The port of Ulladulla today. The tragedy of the *Walter Hood* was keenly felt in this small fishing haven where most of the families have menfolk who work the treacherous offshore waters.

Above: Only the rocks remain... Site of the disaster, with the reef which claimed the fine clipper ship breaking ominously in the background. A memorial to those lost in the tragedy is located on a headland overlooking the reef.

the masts, yards and rigging in a tangled mess onto the decks. The cabin boy was washed overboard and the hull shrieked as timbers snapped and cracked. More huge waves poured aboard and the hull began to disintegrate, spewing cargo into the sea, where broached casks of meat attracted a horde of dark, sinister shapes — sharks!

George Robinson closed his windows and prepared to retire for the night. He went to bed but was restless and tossed uncomfortably for an hour or so. When eventually sleep came it brought the most horrific nightmare he had ever known. He saw a fine clipper ship aground on a storm-swept reef, her crew screaming and drowning before his eyes, while he stood helpless on the shore, unable to do anything to save them. So realistic was this dream that he even recognised the reef as being one a few kilometres along the beach from his property.

Being superstitious by nature, and knowing he would be unable to sleep any more that night, Robinson saddled a horse and rode down to the beach at dawn. As he topped the sand dunes he shuddered with horror. There, before his eyes, was his nightmare — the tortured wreck of the *Walter Hood*. He set off at a fast gallop, this time to the nearest telegraph station to raise the alarm.

By the time help arrived, it was too late for many of the *Walter Hood's* crew. Some had been drowned clinging to the wreck, some had been swept overboard to suffer an even worse fate in the jaws of the sharks. So crazed with blood were these voracious beasts that they attacked even the rescuers who waded into the surf trying to get a line out to the survivors. By the time these exhausted men had been dragged from the sea, daylight revealed the full extent of the tragedy. The beach was littered with debris and bodies. Of the thirty-five crew, only nine had survived.

Eleven bodies were washed up and buried in a communal grave in the sandhills. The nightmare was seemingly over, and the remains of the *Walter Hood* disappeared beneath the waves of the bay that was later named Wreck Bay as a result of the tragedy. But it was not over yet. The wretched bodies of the *Walter Hood* victims were to find no peace in their sandy grave, for wind and water combined to disturb their peace and uncover their remains. A year or two later, a group of men camping on the beach were horrified to find eleven skeletons staring unseeingly at them from the sand. Only when the bodies were removed to a safer place was the nightmare finally laid to rest.

A monument in the coastal bush near Bendalong carries the names of those lost in the *Walter Hood* disaster, and the remains of the ship lie in shallow water near the reef that caused her death. No salvage operations were carried out, since the cargo was considered to be either washed ashore or damaged beyond recovery. What the wreck contains, no one knows, for access to the area is difficult and few divers have explored her remains to any great extent. Perhaps unknown treasures of either monetary or aesthetic value still await adventurers skilled enough to make the dive to her tomb.

CHAPTER EIGHT

HIDDEN BOOTY

THE GOLD COINS OF INVERLOCH

Tucked away in a quiet corner of Gippsland, Andersons Inlet has changed little with time. Surrounded by lush green countryside, it is a natural paradise, virtually untouched by civilisation despite the fact that it is only 150 kilometres from Melbourne. Little disturbs the tranquil surface of its water but the ripples of fish or wildfowl, although occasionally the angry south-westerly wind races in from the open sea, tearing the water into white spindrift, and driving the wildfowl into the shelter of the tall reeds.

When a Scottish farmer, Samuel Anderson, first discovered this inlet, which now bears his name, he was interested only in establishing a port to service the needs of the local community and provide an outlet for the products of his extensive rural holdings in the area. As was the case with many coastal regions in the nineteenth and early twentieth centuries, access by land was virtually impossible and the only communication with the outside world was by sea. Despite the shallow bar across the mouth of the inlet and the unpredictable Bass Strait waters outside, a small port was developed just inside the entrance channel. It was named Inverloch.

The wreath sovereign, circa 1870, minted at the same time as the *Avoca* robbery.
PHOTO ROYAL AUSTRALIAN MINT.

Opposite: A selection of notes and coins which were circulating in Australia between 1840 and 1870.
PHOTO SCOTT CAMERON, COURTESY SPINK AND SON (AUSTRALIA), SYDNEY.

For many years it was a port in name only, for it had no facilities, and supplies for the settlers were brought ashore in rowboats and stored in a shed on the beach until collected by the owners. For some forty years a sturdy little ketch called *Ripple* was Inverloch's link with the outside world, carrying passengers and cargo from Melbourne via Westernport. She provided a spasmodic service, sailing only when she had sufficient bookings, but the passage, albeit somewhat rough, was cheap. The fare for the overnight trip was five shillings (50 c) per head, including meals. Dogs and baggage, however, were carried free.

Like most coastal ports, Inverloch led a sleepy, uneventful existence. The only highlight was the arrival of the *Ripple*, and the only excitement came when the local rocket rescue brigade was called out to a ship stranded on the coast nearby. This was a rare occurrence, which is perhaps just as well, since on one memorable occasion one member of the brigade managed to get the line tangled around his leg and was literally fired overboard when the rocket went off. In the ensuing scramble, the remainder of the brigade were themselves shipwrecked and had to be rescued by the crew of the vessel they had set out to save!

Nothing disturbed the tranquillity of Inverloch until suddenly, in 1877, this quiet coastal settlement was hurled into the public spotlight. Almost overnight it became the centre of one of Australia's greatest maritime mysteries and the focal point for one of the greatest treasure hunts. A treasure hunt that continues to this day, for a cache of gold sovereigns thought to be worth tens of thousands of dollars lies somewhere in the vicinity of Andersons Inlet.

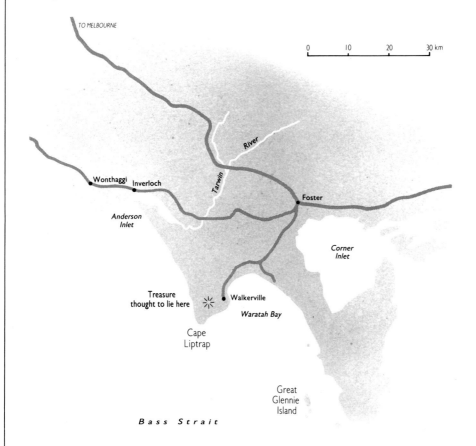

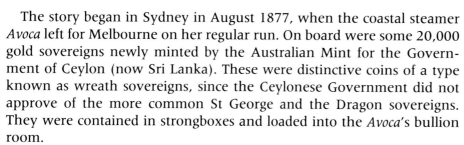

The story began in Sydney in August 1877, when the coastal steamer *Avoca* left for Melbourne on her regular run. On board were some 20,000 gold sovereigns newly minted by the Australian Mint for the Government of Ceylon (now Sri Lanka). These were distinctive coins of a type known as wreath sovereigns, since the Ceylonese Government did not approve of the more common St George and the Dragon sovereigns. They were contained in strongboxes and loaded into the *Avoca*'s bullion room.

The boxes were transferred in Melbourne to the British mail steamer *China* and shipped to Ceylon. But on arrival, one box was found to contain not gold sovereigns but sawdust. Somewhere between Sydney and Ceylon 5000 gold sovereigns had disappeared.

It was obviously an inside job, since the bullion rooms of both ships had not been broken into, nor had there been any tampering with the locks. As the rooms had been opened by a key, suspicion fell on the first mates, who were responsible for the keys. The mate of the *China* was cleared, but not so the mate of the *Avoca*, although he was not charged because of lack of evidence. As the police investigation continued, how-

Map 33
The region around Inverloch and Cape Liptrap where Weinberg is believed to have hidden his treasure.

Top: An original hand-coloured photograph of the old ketch *Ripple* berthing at Inverloch jetty on one of her routine visits.

Fishing boats lie in the shelter of Cape Liptrap. It was here that Weinberg anchored his yacht (perhaps to collect his treasure) when he mysteriously disappeared.

ever, it was discovered that the carpenter of the *Avoca*, a Swedish seaman by the name of Martin Weinberg, had carried out repairs to the door of the bullion room some time earlier. This would have given him an opportunity to make a wax impression of the key without arousing suspicion. Police investigation switched from the first mate to the carpenter, but again there was insufficient evidence for an arrest. The robbery began to take on all the appearances of a perfect crime.

Weinberg left the *Avoca* some months later and took up a selection on the Tarwin River, near Inverloch. Having presumably had enough of sea life, he settled down with his wife and a servant girl, Emma Brackley, to enjoy the pleasant Gippsland surroundings. The furore surrounding the robbery died down and the case was relegated to the files with no further leads or evidence appearing which might give investigating police a breakthrough.

Then one day Emma Brackley was cutting through a bar of soap at the Tarwin property when, to her consternation, a number of gold sovereigns fell out. She informed the police, who searched the Weinberg house, finding a further 40 gold sovereigns in Weinberg's purse and another 35 in a hollowed-out plane. Even more were retrieved from beneath a log at the back of the house. All bore the distinctive wreath impression.

Weinberg was arrested and taken to Melbourne to stand trial. At first he pleaded that a passenger on the *Avoca* had given him the coins, but later confessed to the robbery. In an attempt to lessen his sentence, he offered to reveal the spot where he had hidden a cache of 1700 sovereigns on the Tarwin River. The directions he gave to the hiding place became so involved that police decided to take Weinberg along when they set off to pick up the loot. Inspector Secretan, who had arrested Weinberg, and two detectives made up the party which started out on horseback for the Tarwin River.

At a certain spot on the river, Weinberg told police, there are two trees opposite one another — a blackwood on one side and a tea tree on the other. By stretching a rope between these trees and measuring 32 feet (9.8 metres) from the one nearest his property, they would find the spot where he had dropped an iron kettle, containing 1700 gold sovereigns, into the river. The spot was located easily, a boat was hired from a local boatman, and the search began.

This proved to be harder than expected, for although a heavy object was snagged several times, the police could not get it to the surface. Either it was too heavy or it was itself snagged on something. The grapnels they were using proved too weak, so at dusk a halt was called and the group retired to a nearby hut, commandeered by the police as an operations base. New grappling gear was acquired and at first light the next morning the search began again.

As the inspector and one of the detectives were scrambling into the boat to begin grappling, Weinberg saw his chance. With a hefty blow, he felled the detective who had remained ashore and ran off into the bush. By the time the others had become aware of what had happened and got themselves to the bank, there was no chance of pursuit — Weinberg had disappeared into thick scrub.

Having lived in the area, Weinberg knew his bush, and after a long and fruitless search Inspector Secretan returned to Melbourne to call up reinforcements. For the next few weeks, parties of police searched the bush around the shores of Andersons Inlet, even bringing in black trackers to try and pick up Weinberg's trail. But Weinberg knew how to conceal his tracks by walking along the foreshores at low tide so that the incoming water would wash away his footprints. Although he was sighted on a number of occasions, he always eluded his pursuers.

For five months the police scoured the countryside between Cape Liptrap and Cape Patterson. Living a Ned Kelly style of existence, Weinberg tantalised them by occasionally revealing himself or leaving false trails. There seemed to be no way they could corner him. But, like most clever criminals, Weinberg made one mistake, and it was to prove fatal. On the foreshore of Anderson Inlet, he one day left clear footsteps leading across the sand and into the water. The police were puzzled, for it was unlike the wily Weinberg to leave such an obvious clue. Perhaps he intended them to think he had drowned. Perhaps it was meant to confuse and discourage them. Whatever his reasons, Weinberg misjudged the tenacity of the police. They were leaving nothing to chance. If he had drowned, they would find the body. If he had swum the 2.5 kilometres to the opposite side of the inlet, they would trek round and locate him. Without hesitation, a police party set off to cover the 50 kilometres around the perimeter of the inlet.

Their perspicacity paid off. Weinberg was caught totally by surprise as he lay relaxing on the sand on the far side of the inlet. There was a brief flurry as he jumped over a sand cliff and headed into the bush again, but the police knew they had him cornered. After two shots were fired, Weinberg gave up, and one of Victoria's longest manhunts was over.

But if the manhunt was over, the story was far from over, for there was still Weinberg's trial to be held and a fortune in gold coins to be recovered. In May 1879, Weinberg was brought to Melbourne in the Hastings coach amid incredibly tight security arrangements. Inspector Secretan, no doubt recalling bitterly the escape at Tarwin River, made elaborate arrangements to ensure that his slippery charge would not escape again. The manhunt had created considerable public interest over the weeks and a large crowd had gathered in the city to catch a glimpse of the notorious gold robber.

Concerned that Weinberg might have accomplices in the crowd, Inspector Secretan whisked the fugitive from the coach into a police buggy at St Kilda and took him through Richmond directly to Melbourne gaol. On 30 May Weinberg finally appeared in court charged with the theft of 5000 sovereigns from the bullion room of the *Avoca*.

Repeating his original story that the coins had been given to him by a passenger, Weinberg struck a sympathetic chord in the jury. He was found not guilty of the robbery but guilty of receiving stolen goods, and sentenced to five years in prison. The case was closed, but the coins were still missing. The culprit had been brought to justice, but the mystery was still unsolved.

When he was released from jail, Weinberg disappeared. Despite many rumours and much speculation about his death, no positive confirmation has revealed just how he ended his days. There have been many reported sightings, some in Australia, some in various parts of Europe, but none have been proven nor led to any positive evidence as to the final chapters in the life of Martin Weinberg.

Two mysteries remain to be resolved: how did Weinberg get the gold from the *Avoca* to his property at Inverloch, and where did he hide it? The

JULY 23, 1879.

CENTRAL CRIMINAL COURT

OLD COURT-HOUSE.—TUESDAY, JULY 22.
(Before His Honour the Chief Justice.)
Mr. C. A. Smyth prosecuted on behalf of the Crown.

STEALING GOLD FROM THE S.S. AVOCA.

Martin Weiberg and Joseph Pearce were charged with having, on the high seas stolen 5,000 sovereigns, the property of the Oriental Bank Corporation. They were also in a second count charged with receiving the stolen sovereigns, knowing them to be stolen in two other counts the property was laid as being that of Mr. W. H. Pockley, the captain of the Avoca.

Mr. Purves defended Weiberg; Mr. Quinlan defended Pearce.

The trial was commenced on Monday. The evidence was the same as that adduced at the police court. A clerk from the registrar-general's office produced a copy of the charter of the Oriental Bank Corporation, registered in that office under the act. Evidence was then given that on the 3rd August 1877, a box of gold containing 5,000 sovereigns was sent from the Oriental Bank at Sydney on board the P. and O. Company's steamer, then bound for Melbourne. The gold was put into bags before being placed in the box. The box containing the gold was marked O.B.C. in diamond, 28 under. This box was taken to the office of the P. and O. Company, and was placed in a larger box of the company's, marked XOX. It was packed in sawdust. The outside box was then nailed, sealed, and hoop ironed, put into a truck, taken to the ferry plying between the shore and the Avoca, and placed in the bullion-room on board the Avoca. The sovereigns in this box had been taken by Thomas Gregory, the messenger of the bank, from the treasury. They were what was called the wreath sovereigns, because they had on one side a wreath instead of the St. George and Dragon. The box was consigned to the manager of the company at Ceylon. The key of the bullion-room on board the Avoca was kept by Mr. Elliston, the chief officer. He alone had access to it. The proper entrance to the room was by a hatch in front of the chief officer's room, which was barred and otherwise secured. Since Weiberg's arrest, however, it was discovered that there was another entrance to the orlop deck, on which the bullion-room was situated. This was by a small hatchway aft in the saloon, leading down to the water tanks below in the hold. From this hatchway there was a doorway leading to the orlop deck. On the 7th August, 1877, the Avoca arrived in Melbourne from Sydney, and her cargo was transferred to the China, en route for Galle and on the same afternoon the China sailed for Galle. About the 30th August, 1877, the China arrived at Galle. Two boxes were taken out of the China to be delivered at the P. and O. Company's office. One of them was the large box containing the one in reference to which this prosecution was instituted. The lascars remarked that the box seemed light, and it was taken to the office and there opened. The large box had nothing but sawdust, and the smaller box was partly open, the lid having been prised open. It was empty. Weiberg had been carpenter on board the Avoca, and as such had the right of access to all parts except the bullion room. He remained in the position of carpenter for about five months after the discovery of the loss of the gold, and afterwards took up a selection at the Tarwin River, South Gippsland. Whilst previous living at Williamstown his sister-in-law, Emma Brackley, acted as servant in his house. On one occasion, when she was cutting a bar of soap for washing purposes the knife would not cut the soap. She then got a tomahawk and broke the bar, and found a number of sovereigns in it. On the 18th October, 1878, a warrant was issued for the arrest of Weiberg, and on the 20th Inspector Secretan, Detective Mackay, and Senior-constables O'Meara and Taylor, were on their way to the prisoner's selection to find him. They lost their track and returned to the Powlett river bridge, where they baited their horses. Whilst stopping at a house the prisoner rode up. Mackay said to him they were on their way to the Tarwin, but had lost the track, and that they were going to select some land. Weiberg said, " I have just come from there. I shall be back in a week, and shall be most happy to show you round." Mackay asked him whom he should inquire for, and he said his name was Weiberg. Prisoner was then arrested on a charge of stealing the 5,000 sovereigns from the Avoca. A purse was found upon him ... sovereigns, and in ...

first will probably never be known. It seems highly unlikely that 5000 sovereigns could have been taken off the ship, through the dock gates, and by road to Inverloch without arousing suspicion. Perhaps an accomplice in a fishing boat picked up the coins, thrown overboard from the *Avoca* in a floating container as the ship passed Anderson Inlet.

More important of the two mysteries is the second — the location of the treasure. There have been numerous searches over the years, but not even a single gold sovereign has come to light. Weinberg's hut stood on a property owned by a Mr Wyeth, of Pine Lodge, Inverloch. When it was pulled down in 1940, the remains were thoroughly searched but no gold was found. Knowing that police would continue looking for it, Weinberg is sure to have found a safe hiding place. This would have been easy because of his considerable knowledge of the area. Most likely it is buried underground, or under water, somewhere in the area where he was a fugitive for five months.

Over the years, a number of clues have come to light which would seem to focus the likely position of the treasure in the Cape Liptrap area rather than near the Tarwin River or Inverloch. The strongest came as a result of the most reliable sighting of Weinberg after he was released from prison. The manager of the lime kilns at Waratah, a Mr Dewar, claims to have spoken at length to Weinberg, who identified himself quite openly. It appears that Weinberg had purchased a yacht which he had anchored under an island some distance away, probably near Walkerville, under the lee of Cape Liptrap. Had Weinberg returned to collect his treasure?

This line of thought is reinforced by the fact that later, after Weinberg was believed to have drowned, his two brothers arrived at Waratah and made no secret of the fact that they were searching for the treasure. They spent countless days combing the bush in the vicinity until one of them was killed in a fall. Since they had been with Weinberg on the yacht just before he was supposed to have drowned, the inference must be that he had indicated the location of the treasure, but had not given them specific details. This again would seem to support the theory that the treasure lies somewhere on or near the promontory of Cape Liptrap.

TREASURE ON MOUNT WHEOGO

Mountain ranges always make good treasure-hunting country, and the Great Dividing Range, which forms a spinal column down the entire east coast of Australia, is no exception. The crustal upheaval that created this massive range millions of years ago brought close to the surface the hidden minerals that were contained in older rocks deep below the surface. The weathering and erosion that followed cleared away the overburden, exposing deposits of brilliant gemstones or glittering seams of gold.

But although gemstones are found in profusion in the rugged central ridges of the Great Dividing Range, most gold strikes of any consequence have been made on the gentler undulating country known as the Western Slopes. This has always been a prosperous region. Wheat, sheep and cattle all thrive on the pastures watered by the big western rivers, and the scattered towns are a constant buzz of rural activity. But never has the region been so prosperous as in the middle years of the nineteenth century when gold fever struck.

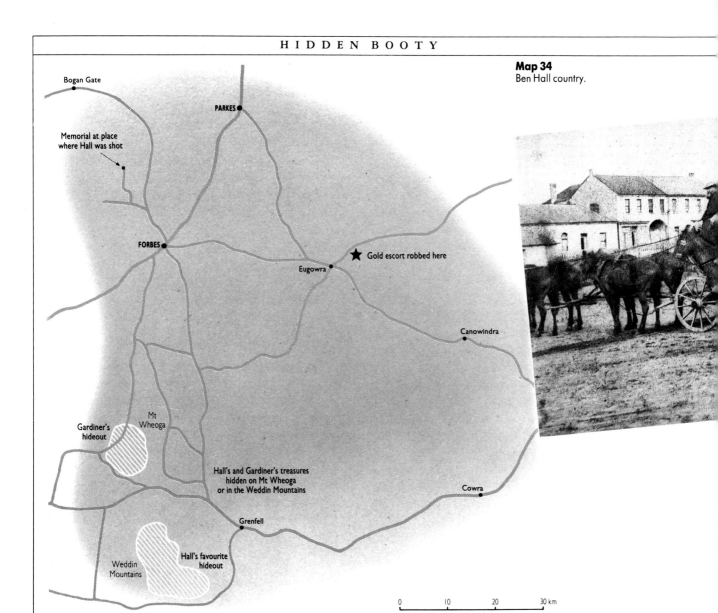

Map 34
Ben Hall country.

Bogan Gate

PARKES

Memorial at place
where Hall was shot

FORBES

Eugowra

★ Gold escort robbed here

Canowindra

Mt
Wheoga

Gardiner's
hideout

Hall's and Gardiner's treasures
hidden on Mt Wheoga
or in the Weddin Mountains

Cowra

Grenfell

Weddin
Mountains

Hall's favourite
hideout

0   10   20   30 km

Within a year, the fertile soil was torn apart in the frantic efforts to see what lay beneath it. Gold not only seduced sane men from their families, city men from their offices and farmers from their land, it was also the catalyst that ushered in a dramatic and unique chapter in Australia's history — the era of the bushranger.

Mountain ranges make good bushranging country, not so much for the treasures they contain — for bushrangers were notoriously lethargic in their methods of obtaining wealth — as for the way in which the environment lends itself to their unlawful occupation. In the early bush-ranging days, coach roads were little more than narrow tracks hacked from the mountainsides, creating a vulnerable situation for the already vulnerable coaches. Huge overhanging boulders provided ideal opportunities for an ambush, and there was no better cover for a getaway than the dense scrub that blankets most of the ranges.

Although bushrangers were not unknown before the gold rush, they were mostly footpads or petty criminals, striking at isolated targets and gleaning small pickings. With the discovery of gold, however, the prospect of large takings lured a more organised and more daring type of

Grenfell about the time Gardiner and Hall operated in the area (left). Timber-built country banks were very vulnerable to attack by bushrangers. The gold escort at Bathurst in 1872 (right). This one made it in from the goldfields without interference from bushrangers!

MITCHELL LIBRARY, STATE LIBRARY OF NSW.

criminal into the business. Lack of security in remote areas meant easy plunder for these bushrangers, who mostly ran in gangs and attacked coaches carrying gold from the fields to Sydney or other main centres.

One of the most sophisticated of the bushrangers at this time was Ben Hall. Although intelligent and cunning, Hall at the same time revealed the sympathetic and benevolent nature of a latter-day Robin Hood. He is believed never to have killed a man, despite extreme provocation.

Hall was born in Australia in 1837 and was involved in a few minor skirmishes with the law before turning to bushranging in the gold-rush days of the 1850s. At first he was content to be a partner in a gang led by the notorious Frank Gardiner, but his natural leadership led him to form his own gang when Gardiner moved to Queensland. Before they broke up, the partners planned their biggest coup together. It was to be the most daring and intrepid exploit of their unlawful association — they would rob the gold escort!

Parcels of gold brought into Forbes from the surrounding diggings were loaded aboard a special coach or wagon and conveyed to Bathurst under police escort. There had never been a robbery involving escorted gold, so security arrangements had become somewhat lax. The gold escort left Forbes every Sunday at exactly the same time and followed exactly the same route. There was no mounted escort, and of the four troopers who accompanied the gold, only one sat on the box-seat with the driver, the other three remaining inside.

Gardiner and Hall planned the robbery with all the precision of a military operation. Gathering together six hand-picked men, they selected Sunday, 15 June 1862, as the date and a spot on the road about 43 kilometres from Forbes, just past the little town of Eugowra, as the location. The site was an uninhabited stretch of road with a modest uphill gradient so that the coach would be travelling slowly, with no chance of spurring the horses to break through the ambush. A large rock at the spot selected for the robbery now carries a plaque commemorating the events of that fateful day.

The men gathered at the scene early in the morning to begin their preparations, for nothing was to be left to chance. Two cumbersome bullock teams that made their way laboriously up the hill were immediately commandeered by Gardiner as useful acquisitions for the holdup. Swung broadside across the road, they created a very effective road block. With a boulder-strewn hillside above them and a steep gradient below, the road was effectively sealed and the bushrangers settled down to await the arrival of the coach.

The road block was not only effective, it was cleverly disguised, for as the police approached they saw nothing suspicious in what appeared to be a couple of drunken bullockies who had lost control of their teams. Cursing loudly, they swung the escort coach to the side of the road and attempted to drive around them. The coach lurched unsteadily off the shoulder of the road as the driver coaxed his horses past the first bullocks. At that moment the bushrangers, their faces covered, sprang the trap. Suddenly aware of what was happening, the trooper on the box seat reached for his rifle, but was felled by a shot from behind the surrounding rocks. Another volley greeted the three troopers who tumbled from the coach, wounding one and causing the others to take to their heels. A third volley frightened the already terrified horses which reared and plunged, tipping the coach onto its side. The bushrangers moved in quickly and unloaded an estimated £14,000 in gold and cash.

It was all over in a few minutes. The bullockies, paid handsomely for their trouble with a pocketful of gold coins, continued their leisurely journey as though nothing had happened. Gardiner, Hall and their gang loaded the loot onto packhorses and headed westwards towards their hideout at Mount Wheogo. The police, with two of their number wounded, lay low in the bush and waited for assistance. For the first time in Australian history, a gold escort had been robbed!

At their well-established hideout on Mount Wheogo, some 70 kilometres from the scene of the ambush, the gang met to divide the spoils, each receiving around £1,750 in gold and cash — a princely sum for those days. They split up quickly, but not quickly enough. A police patrol, following hard on their trail, spotted four of them and gave chase. Hindered by their packhorse, heavily laden with the spoils of the robbery, the outlaws could not shake off their pursuers and eventually had to cut loose the exhausted beast. Much to their chagrin, it was found by the police and its precious cargo confiscated. The gang had bought their freedom with half the proceeds of the robbery.

In the intense manhunt that followed, police combed Mount Wheogo, the nearby Weddin Mountain Range and much of the surrounding countryside, but without success. Some weeks later, however, when the hue and cry seemed to have died down, two of the gang were caught heading for the Victorian border. With them was their share of the escort robbery, some £3,500. In a daring ambush, a few hours later, Frank Gardiner managed to rescue his men but could not recover the gold or money.

New South Wales was getting too hot for Gardiner, so he headed north for Queensland. Ben Hall, who at this stage had not been identified as one of the escort gang, returned to his small farm, having first hidden his share of the spoils in a waterhole. It is quite possible that, left to his own devices, Hall would have resumed his life as a law-abiding citizen, but fate decreed otherwise. Inspector Pottinger, the man in charge of the

The exact spot where the escort robbery took place. Tracks of the old coach road are visibly etched into the rock in the foreground, while the large rock in the distance provided cover for the ambush gang.

The Weddin Mountains, near Grenfell. Ben Hall's hoard is thought to be concealed in this mountain range.

police search, had desperately tried to indict Hall, but without success. In a fit of rage, he one day set fire to Hall's farm and destroyed everything, including his stock, leaving Hall with neither a home nor an income.

Bitterly swearing revenge on Pottinger, Ben Hall from that moment turned his back on the law. Joining forces with John O'Meally, another member of the gang, Hall embarked on his career as Australia's most notorious bushranger. The two men holed-up in a cave some 30 kilometres from Mount Wheogo, carrying out a series of daring sorties in the surrounding district, much to the chagrin of Inspector Pottinger who had sworn to bring them to justice. But Pottinger was soon to rue the day he began his personal vendetta against Hall, for the bushranger was to make the vindictive policeman the laughing stock of the nation.

It happened one day when Pottinger's men cornered Hall in a patch of scrub. The police dismounted and fanned out to surround the area, moving in slowly and tightening the cordon. Hall also dismounted, and with a slap on the rump, sent his horse off at a gallop, while he climbed a nearby tree. The police, hearing the horse dash through the scrub and assuming the bushranger was trying to break out, ran to cut him off. Hall dropped from his tree, selected the best of the police horses — a fine thoroughbred racehorse — and sent the others in all directions.

The horse he had selected was the personal pride and property of none other than Inspector Pottinger. When the footsore police patrol returned to their station minus their horses, they faced the ridicule of their comrades as well as that of the public. The story made excellent copy for the newspapers, which gave it headline treatment right across the nation!

Now Hall's bushranging exploits begame legend. He cleverly planned robberies of banks, stores, hotels and coaches, each time changing his tactics and each time striking in a different area. He established a number of different hideouts, all well stocked with ammunition and food, never using any one for more than a few days at a time. For a number of years the audacious bushranger led the police a merry dance, always confidently calling his own tune and always successfully disappearing without trace.

So daring were some of his activities, and so unrewarding the results, that it seemed Hall was as much attracted by the thrills and excitement of the chase as he was by the financial gain. Or perhaps the real stimulus was a fierce desire to humble his sworn enemy, Inspector Pottinger. By this time Hall's successes and the failure of the police to apprehend him were being given wide coverage in the press, and the hapless Pottinger was rapidly becoming a favourite subject for ridicule and lampoon.

No matter what the reason behind his lawlessness, there was no question that Hall was building a substantial fortune. In typically audacious fashion, he even opened a bank account, through a friend, and deposited £6,000, much of which was probably the result of earlier bank robberies.

But the law was beginning to catch up with bushrangers generally. In the five years between 1862 and 1867, twenty-three bushrangers were shot dead or hanged. In 1865 a new law was passed by the New South Wales Government under which all bushrangers were to be shot on sight if they did not immediately surrender. In addition, those who harboured them were liable to 15 years imprisonment plus confiscation of all their goods and property.

The pressure was beginning to tell on Hall. He was now 28 years old

Hall was caught sleeping beneath the trees in the middle distance, and shot dead before he could escape.

Ben Hall's grave in Forbes cemetery.

and a wealthy man. In May 1865 he decided to give up his hazardous existence and settle for a life of ease and luxury. Just where he would go and what he would do was a problem, for Hall was now a prince among outlaws and his face was known to every policeman in the land.

It is doubtful if the police would ever have caught up with him if they had been forced to rely on their own efforts. But the new law, with its threat of dire punishment for accomplices and generous rewards for informers, was the catalyst which finally caused Hall's downfall. He was betrayed by a friend who, fearful of losing his property as the result of his association with the bushranger, led police to his hideout.

The end came swiftly. Surprised while he was asleep, Hall was shot once before he had time to reach for his gun. As he sank to the ground, the police moved in, firing twenty-seven bullets into his body. It was mercifully quick and, had he been aware of it, he would have been gratified to know that at the end his nemesis was not the hated Inspector Pottinger, but Inspector Davidson of the Forbes police.

Both Hall and Gardiner are known to have buried much of their booty in the bush of the Weddin or Wheogo Ranges. There is, of course, no documentation on either the amounts or the locations of this hidden treasure, although such evidence as is available points convincingly to its existence.

Gardiner, who lived out his life in California, is reputed to have written to an associate in the Grenfell area enclosing a rough map of the spot where his cache was hidden. The man, known as Fogg, searched spasmodically for the rest of his life, but without success. What eventually became of the map is not known.

Many years later, two American men arrived in the area, ostensibly to prospect for radium. They stayed for some time and concentrated their efforts in the Mount Wheogo area, talking to no one and offering no clue to their real intentions. When they left, it was discovered that they had dug up the ground to a considerable depth around the exact spot where Gardiner had his hideout and where the proceeds of the gold escort robbery were divided up.

The identity of the two men was never revealed. There is speculation that they might have been Gardiner's sons, or two acquaintances he had made in California. Perhaps he had divulged the secret to them, or perhaps they were just treasure seekers acting on clues they had picked up in conversation with him. Whatever their background, their information must have been well-founded, for they knew the exact location of the bushrangers' hideout.

Hall is known to have secreted his share of the escort robbery in a waterhole in the same area. Of the vast amount he accumulated from his later activities, only the £6,000 in his bank account was recovered, so it is safe to assume the remainder lies hidden somewhere in the same mountains. His son, Harry, searched for years but was unable to locate either gold or cash.

So, as far as the evidence points, the caches of gold and coin hidden by both Ben Hall and Frank Gardiner still lie somewhere in the timber-covered mountain ranges near Grenfell. They could be in the Weddin Mountains, now partly national park and partly state forest. But most pointers indicate that Mount Wheogo should be the starting point of any search. Ben Hall's son Harry, Gardiner's associate Fogg, and the two mysterious Americans all focused their activities on this mountain to the north-west of Grenfell. These are the only people thought to·have any positive information of the whereabouts of the treasure. They did not find it, and despite numerous searches since, neither has anyone else.

QUEENSCLIFF'S DARK SECRET

The little town of Queenscliff sits on a promontory of the Bellarine Peninsula overlooking the turbulent waters of the Port Phillip Bay "Rip". Queenscliff has had a chequered history since a few fishermen's huts were first built there in 1847. Fishing is still its main industry, although the combination of still waters on one side of the peninsula and ocean surf on the other, together with the fact that it is only about 100 kilometres from Melbourne, have now made it a popular holiday resort.

But it is its location at the narrow entrance to Port Phillip Bay which has dictated its real role — that of sentinel at the gates of Australia's second largest city. Every ship that has entered Port Phillip Bay since 1839 has been checked at the Queenscliff pilot base station and provided with a sea pilot if required. To this day the identity of every ship that enters or leaves the bay is recorded, and for all but a few local vessels, the use of a pilot is compulsory.

Like the heads of Sydney Harbour, Queenscliff and its companion town of Lonsdale Point have provided the strategic forward defence locations for protection against attack from the sea. Barracks and military fortifications were erected in the 1850s when, at the height of the Crimean war, a Russian invasion was expected. Fort Queenscliff was built in 1875 and for over 70 years provided the main fortification for Port Phillip Bay. With modern war technology, such forts became obsolete, and in 1947 the fort was converted to an Officer Training School.

Although there was a very realistic scare of a Russian invasion in 1850, and all coastal forces were put on red alert, it turned out to be a newspaper hoax, so the cannons at Queenscliff were never fired in anger and the historic little town was never the scene of violent confrontation. Indeed, most of the excitement in the area has been involved with shipwrecks, of which there have been many, both in the treacherous Rip, and along the nearby coastline of Bass Strait.

But some of the town's more interesting historical events may have been played out long before Queenscliff was even a name on the map, perhaps even before the first convicts arrived in Port Jackson. There are legends and stories, some with quite convincing evidence, of maritime visitors to the area in the days when the east coast of Australia was known only to the Aborigines.

In 1847, a bunch of five keys was uncovered at an excavation in Geelong, some 30 kilometres away. The location, age and archaeological evidence surrounding this find indicate that these keys could have belonged to the Portuguese explorer Cristovao de Mendonca, who is thought to have visited the area in 1522. The discovery and location of the "Mahogany Ship" (see Chapter 9) adds further weight to the fact that the Portuguese sailed along the coastline long before any British, Dutch or French navigators arrived.

If Mendonca visited Port Phillip Bay in the sixteenth century, then it is possible that other seamen followed his lead. Perhaps not so much with

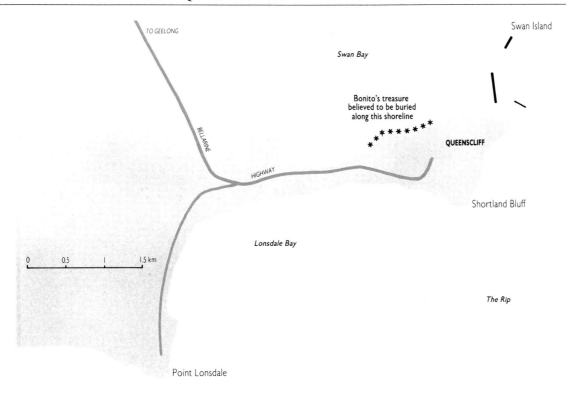

Swan Island

Swan Bay

Bonito's treasure
believed to be buried
along this shoreline

✴✴✴✴✴✴✴✴
✴   ✴

QUEENSCLIFF

TO GEELONG

BELLARINE

HIGHWAY

Shortland Bluff

Lonsdale Bay

The Rip

0   0.5   1   1.5 km

Point Lonsdale

Map 35
The location of Bonito's hidden
treasure.

a view to settlement as for other reasons. Whalers, who spent much of their time in the inhospitable waters of the Southern Ocean, often holed-up in sheltered bays and inlets on the Australian coast to rest their crews and repair their ships. And an unexplored coastline must have been extremely attractive as a hideout for pirates and other lawless mariners.

One such pirate was Benito Bonito, about whom many tales have been woven, probably most of them apocryphal, since no evidence to substantiate the stories has ever been produced. However, Bonito did exist and was a well-known pirate in his day.

It appears that Bonito somehow came into possession of a very large and very valuable cargo of gold and precious stones. Such was his reputation that the only possible source of such riches could have been a plundered ship or a ransacked city. One story connects it to Peru, where, in the middle of an uprising in the early nineteenth century, the valuable possessions of Lima city, particularly those of its cathedral, were shipped off to an unknown destination for safe keeping. Bonito attacked the ship and stole the treasure, then sailed west in search of a safe hiding place.

Two locations for this hiding place have been pinpointed — the Cocos Islands and Queenscliff. Although there appears to be no documentary evidence to suggest either the truth of the story or the possible location, a number of well-organised and heavily financed expeditions have carried out extensive searches on the Cocos Islands. One such expedition was mounted at huge expense by the British adventurer and speed ace, Sir Donald Campbell. Another was organised by experienced veteran treasure hunter Harry E. Rieseberg. A recognised authority on recovering lost treasures, Rieseberg had pinpointed and recovered many sunken treasures before attempting to find Bonito's cache. Another man spent 20 years digging up the Cocos Islands before deciding that the treasure was quite definitely not there!

Meanwhile, the Queenscliff site was also coming in for some attention. The story here was that Bonito landed at Queenscliff in search of a site to hide his loot. A cave on the Port Phillip side of the peninsula attracted his attention, but while unloading the treasure and transporting it to the cave, the pirates were surprised by a British warship. To conceal the treasure, they brought down the mouth of the cave with a charge of gunpowder. They were no match for the warship and shortly afterwards the pirates, including Bonito, were captured and later hanged, taking the secret of the treasure with them to the gallows.

Although there was no documentary evidence of any kind, numerous treasure seekers attempted to find Bonito's cache. The first serious attempt was in 1937 when a syndicate from Daylesford in Victoria hired the services of a metal diviner. This was in the days before metal detectors, and the diviner worked with a rod, rather along the lines of the water diviners. Soon the diviner claimed he had located a large mass of metal beneath the surface. It was near the shore in roughly the spot where Bonito's treasure was thought to be, so the expedition quickly got to work.

First they dug a shaft on the precise spot indicated by the diviner. At around 15 metres, the shaft revealed an underground cave, but because it was below sea level, they were unable to search the cave. Huge pumps were brought in but could not stem the flooding of the water, so a diver was sent down with powerful underwater floodlights. Sand was excavated from the cave and sent up by bucket in an effort to locate anything lying buried underneath, but the continued flooding of the water and subsidence of the sand finally brought the work to a halt.

Two years later, the syndicate, refinanced and reorganised, with far more sophisticated equipment, returned to the site. Steel cylinders in which a man could work were lowered into the cave in an attempt to remove the sand, but again the treasure hunters were thwarted by the water and quicksand. The attempt had to be abandoned, but members of the expedition remained emphatic that they had found the treasure.

Since then, numerous other attempts have been made to recover Bonito's booty, but all without success. There is, of course, nothing to prove that it is there, but, as any true treasure hunter will always counter — there is nothing to prove that it is *not* there!

THE LONE (BUSH) RANGER

It is not hard to believe that wherever there was a bushranger, there could also be hidden treasure. The rogues that plagued the inland areas of the colony in the mid to late nineteenth century reaped huge hoards of wealth from their victims, particularly during the era when gold was carried about the countryside in pockets, purses and hand luggage. The hold-up of a stage coach on its way back from the goldfields could return a fine haul for a morning's work, and when the more daring and well-organised gangs bailed up the bank in a gold town, or the gold escort on the way to the city, they usually rode away with undreamed-of riches.

But their lifestyle did not provide the bushrangers with much opportunity to spend their loot. Nor could they stash it in a bank, as Ben Hall found out to his cost. Brazenly banking the proceeds of what prob-

Fred Ward (Captain Thunderbolt),
photographed after his death.
MITCHELL LIBRARY, STATE LIBRARY OF NSW.

ably included the takings of bank robberies, Hall built up a nice nest egg in a New South Wales country bank. But when he was betrayed, his bank account came to light and his savings were confiscated.

So most of the proceeds had to be cached around the countryside until such time as the bushranger retired from his hazardous business. But retirement became harder to achieve as the police stepped up their campaign against the outlaws, and few had the opportunity to enjoy the proceeds of their years outside the law. Gaol sentences were long, and the gaols were in such primitive condition that even those who escaped the gallows rarely survived a long internment behind bars.

As a result, many hidden caches of bushrangers' spoils must still lie concealed in the mountains and the bush. Only the few who were eventually released from gaol would have had the opportunity to retrieve their hoard. In all probability they were marked men and dared not go near their former territory for fear of arousing the curiosity of the police.

One bushranger who met his end before he had time to enjoy his spoils was Fred Ward, who roamed the New England Tableland preying on travellers passing through the region between Tamworth and the Queensland border. Better known as Thunderbolt or Captain Thunderbolt, Ward spent five years in his illicit trade, during which time he must have amassed a sizeable fortune.

Since he was shot dead in 1870 while still in his prime, Ward must have left his loot stashed away somewhere in the area he knew so well. Although he was known and liked by the local folk, he did not mix socially and preferred his own company at most times. Unlike most other bushrangers, he operated alone, so there was no one else to share the secret of his hidden hoard. He had no family, and his long-time lover died three years before him.

Ward was the typical lone ranger. His exploits were almost always carried out single-handed, and he had a dashing, troubadour style about him that appealed even to his victims. After a hold-up he would often ride away singing at the top of his voice.

Once he is reputed to have bailed-up a coach carrying a German brass band on their way to a performance at Tenterfield. First he made them play a tune and then took their money. When the leader of the band protested that the musicians would be stranded with nothing, Thunderbolt apologised but said he needed the cash to put on a "dead cert" at the Tamworth races next day. As he rode off in cavalier style, he promised that if the horse won, he would return their money. The horse did win and the Germans recovered their money at Tenterfield post office!

As time went by, Thunderbolt became a latter-day Robin Hood, often helping local people with charitable deeds and sympathetic understanding. His victims were rarely critical of his behaviour. Even though they lost their possessions, they never had to fear for their lives, for the outlaw of the tablelands was always courteous and, as far as is known, never harmed his victims. This reputation stood him in good stead, for when police patrols were investigating his activities, they received little or no co-operation from locals.

So concerned were the authorities at Thunderbolt's activities that when banknotes were sent by coach through the area, the notes were cut in half and sent in separate consignments. This indicates the degree to which the bushranger's operations had succeeded and, by inference, the huge amount of wealth he must have amassed. Always riding a thoroughbred mount, and always planning his robberies meticulously, Thunderbolt was rarely in danger of being captured, even when the police closed in on him. Laughing and singing, he rode off into the countryside he knew so well, leaving the police frustrated and angry in his wake.

A natural horseman, he would often lead his pursuers to a spot where he could lose them simply by plunging deep into dense bush, or leaping across a chasm which he knew the police horses would refuse. One such chasm is called Thunderbolt's Leap and has become a part of New England's colourful history.

Then, one day in 1870, the Lone Ranger of the New England Table-lands met his end. After a long chase through the bush near Uralla, Fred Ward, alias Captain Thunderbolt, was cornered. After a brief gun duel the outlaw was mortally wounded with a bullet in his chest. He died before police could reach him. He is buried in Uralla cemetery where to this day his grave, with a headstone erected by friends, attracts considerable attention.

Just where, among the peaks and valleys of his beloved New England Ranges, Thunderbolt stowed his ill-gotten gains, no one knows. History does not record any details that might lead to the recovery of his loot — a factor which would have been very prominent in police investigations. Nor has there been a discovery of hidden treasure in the area which could be attributed to him.

Somewhere among the bitterly cold mountain tops, or perhaps in a cave which once gave him shelter, lies the fortune of Captain Thunderbolt. Perhaps one day some enthusiastic treasure hunter will track it down and put to rest for once and all, the mystery of Thunderbolt's treasure.

Thunderbolt's grave at Uralla.

Typical Thunderbolt country. The big rock formations, known as tors, provided good cover for marauding bushrangers.

CAPTAIN MIDNIGHT'S HOARD

Like the actors that many of them were, the swashbuckling bush-rangers who roamed the countryside in the latter part of the nineteenth century adopted stage names by way of feeding their egos and, possibly, terrorising their victims. The very mention of Captain Thunderbolt, Mad Morgan, the Wild Scotchman, Black Douglas or The Stripper was enough to strike terror into the hearts of timid damsels and send recalcitrant children scuttling to bed, pulling the covers firmly over their heads. Even Rolf Boldrewood, in his classic Australian novel *Robbery Under Arms*, gave his bushranger the name of Captain Starlight.

There was obviously no register of proprietary names among the bushranging folk, for in 1878 no less than two bushrangers were opera-ting under the title of Captain Midnight. The Victorian version, Andrew George Scott, later became known as "Captain Moonlite", possibly to avoid confusion, while Alexander Law, alias Henry Wilson, alias George

Map 37
Probable location of Captain Midnight's hoard.

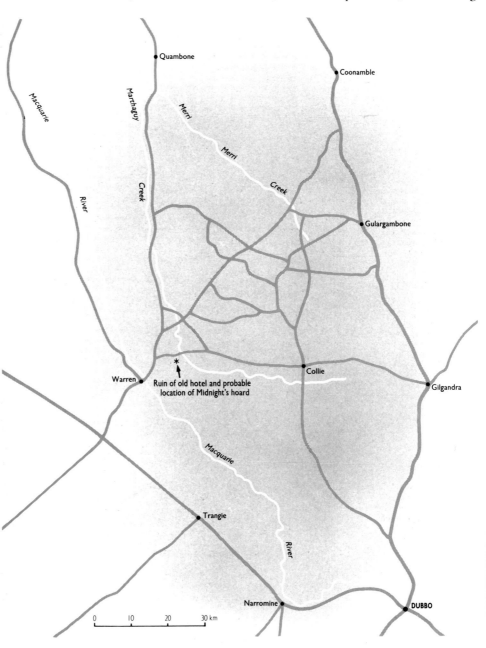

Gibson, who roamed the western slopes of New South Wales, retained his claim to the title of "Captain Midnight". Both were desperate outlaws and both died as the result of killing a policeman, and no doubt both had loot hidden somewhere in their territory. But the Victorian roamed a large area of his native state, and also entered southern New South Wales, whereas the activities of Captain Midnight were confined to a fairly limited region, so the possible location of his hidden hoard can be more easily pinpointed.

Law, alias Wilson, alias Gibson, alias Captain Midnight was an escaped convict who roamed the Dubbo–Wellington region of the central west of New South Wales in the early 1870s. His activities revolved mostly around stealing horses in the Merri-Merri and Marthaguy districts, and occasionally bailing-up and robbing a stage coach or some wealthy traveller. Life was tough in the country in those days and since Midnight was not known for violence, he was tolerated by the locals, many of whom had themselves fallen on hard times at one time or another and had to move outside the law to survive.

Where the coach road crossed Marthaguy Creek on its way north, an enterprising lady called Jane Flick selected forty acres, opposite Wonbobbie Station, and built a hotel called the Travellers Rest. It was a perfect site for an inn and became a popular stopover for mail coach passengers. It was also popular with the locals, and in particular with a gang of horse thieves who worked with Captain Midnight. Well removed from either Dubbo or Wellington, the gang were never troubled by police and became quite brazen about using the hotel openly. It would be safe to assume that this was Midnight's headquarters in which many a robbery or hold-up was planned.

But on Friday, 20 September 1878, the bushranger's fortunes took a turn for the worse. At six in the morning, Senior Sergeant Thomas Wellings, of the Dubbo police, together with Senior Constable Souter and Constable Walsh, moved in on the Travellers Rest. They had been notified by Sydney that an escaped convict from Parramatta gaol was in the vicinity. Although the man they sought was not in the hotel, Captain Midnight was. As Wellings entered the pub, Midnight ran out of a back door and into the bush. He was spotted by Constable Walsh and the chase was on.

Cornered in a paddock, the bushranger turned and kneeled. Taking careful aim, he shot Sergeant Wellings dead and wounded Constable Walsh's horse, causing it to bolt. Midnight disappeared into the bush and shortly afterwards stole a magnificent thoroughbred horse on which to make his getaway. But having killed a policeman, as many a bushranger found to his cost, his days were numbered, and a huge police search began across the central regions of the state.

On 2 October, Midnight was seen in a village near Bourke and challenged by a police patrol. He was riding the stolen thoroughbred and leading two well-laden packhorses, probably intending to make a new life across the Queensland border. Abandoning the packhorses, which also turned out to be thoroughbreds, Midnight took off with the police in pursuit, and shook them off by plunging into a lignum swamp. But the police were not going to be so easily evaded, and black trackers were brought in to follow his trail.

Two days later, Captain Midnight was found asleep under a tree on Maranoa Station, close to the Queensland–New South Wales border. In

A typical scene in the central west of New South Wales, where Captain Midnight roamed.

a desperate attempt to escape, he leapt onto the thoroughbred, but a fusillade of shots killed the horse instantly and mortally wounded the bushranger. He was carried to a nearby hut on Wapweelah Station where he died the next morning.

Midnight was known to have amassed a sizeable hoard of loot from his years as a bushranger and rumour has it that he hid it in one of the yards on Wonbobbie Station or near the Travellers Rest hotel. However, the station had numerous yards in 1878, and which one was favoured by the bushranger is not known. The hotel has long since gone but the cellar cavity near the creek can still be seen, as can posts from nearby yards.

Midnight told no one of the exact location of his cache, so the secret, like so many secrets in treasure tales, went with him to the grave.

CHAPTER NINE

LOST TREASURES OF
LEGEND

There are a variety of reasons for hiding treasure. Some people do not trust banks and so conceal their valuables in places where thieves and robbers are unlikely to find them. Illicit goods, or those obtained by questionable means, are often concealed until they can safely be brought out into the open.

In Australia's frontier days, banks were notoriously unreliable, and thieves and robbers were legion, so any items of value were concealed in a secret hiding place known only to the person who hid them. Bushrangers, with the police hot on their tails, hid their booty until the danger was past. Old prospectors, striking a rich deposit of gold, told no one until they had the means to exploit it fully themselves.

This is not an uncommon human trait — to protect your personal treasure by throwing a cloak of secrecy around it. But a problem arises when the only person who knows the secret dies. Then, as often as not, the secret dies with him. However, an equally natural human trait is the need to talk about a secret, and most of those who hid their treasures left some clue to its whereabouts. A casual comment to a friend, a diary with a few details, perhaps even the traditional treasure map, with "X" marking the spot.

However slight the clue, gossip and speculation soon enlarges it until the story of the hidden treasure becomes a legend. The history of Australia is awash with such legends, most concerning hidden riches and mystery treasures, many based on fact, some cultivated from hearsay. Knowing which is authentic and which is not creates a problem for would-be treasure seekers. Careful research can dispel some of the myths, but since few hidden treasures are documented, research often does not authenticate the legend.

Many stories of hidden treasure have been handed down through generations of Aborigines and Islanders, where fact often has been mixed with tribal legend or superstition. Similarly, old bushmen and swaggies, swapping tales around the camp fire, soon manipulate fiction into fact or reality into myth. Sorting fact from fairytale is the hardest part of determining whether a story of hidden treasure should be followed up. None of the stories that follow can be vouched for, but all have the potential to be true. There can be no guarantees that the treasures even exist, but if they do, some lucky treasure hunter will one day reap the reward.

Opposite: Any serious seeker after lost treasures might need these tools of the trade.
PHOTOS SCOTT CAMERON.

157

CRANGER'S REEF

The area around Coen, in the central regions of the Cape York Peninsula, has always been gold country. To the south is the famous Palmer River, where some of Queensland's largest gold strikes were made. Close at hand are mining districts with such famous names as Ebagoola, Klondyke, Rocky River and Choc-a-Block. Good finds have been recorded at Coen itself, a dusty settlement on the Peninsula Development Road.

In 1872, when the Palmer River goldfield was in full production, a few prospectors were pushing north in the hope of finding more rich deposits in the jungle-clad hills near the coast. Mottler's store, the only supply centre in the area, was doing a roaring trade and, like the local pub, was a focal point for both the locals and the prospectors working the area. Jim Mottler knew most of them by name and shared the ups and downs of their fluctuating fortunes. One day an old prospector by the name of Frank Cranger staggered into Mottler's store and threw onto the counter a handful of gold. He told of a reef richer than anyone had ever seen before which he had discovered in the thick, almost impenetrable jungle of the nearby McIlwraith Ranges. Gathering up his supplies and promising to be back soon with more gold, Cranger plunged back into the dense vegetation, never to return.

Mottler found his body, riddled with spears, close to the old-timer's base camp. There was no sign of gold nearby and, despite an intensive search of the area, Mottler could not find Cranger's gold reef or locate any tracks that might lead him to the treasure. He returned to his store convinced that whatever Cranger had found, the secret of its location had died with him.

But the story of the mystery reef had become known in the area and it was not long before other prospectors began to take up the search. At the time, the hostility of the local Aborigines was at its peak and white men were being killed almost daily. Despite Mottler's warnings, two local prospectors loaded themselves with supplies and set off into the jungle to reach Cranger's base camp and, from there, to search for the reef. They found the camp but were speared to death before they had time to examine even the immediate area.

More treasure seekers followed. Soon they were arriving in hundreds. Lured by the story of Cranger's gold, they ignored the dangers and smashed their way into the dense undergrowth, searching every ridge and every gully in an effort to find the fabled reef. Some were killed by blacks, some suffered terrible hardship, some almost starved to death. One party, after losing its way in the thick vegetation, wandered more than 240 kilometres across the Cape and were found by timber cutters on the shores of the Gulf of Carpentaria.

For twenty years, men came from the goldfields all over Queensland to search the rugged country around Coen. Then the hunt took a fresh turn when two prospectors announced that they had found new clues to the whereabouts of the elusive reef. Fully equipped with supplies from Mottler's store, they headed into the bush, following their new lead until they reached a swamp. Unbeknown to them, they were, in fact, quite close to Cranger's reef, but once more fate stepped in and defeated them.

One of them, a man named Mellone, was attacked by a crocodile, which ripped away a large part of his leg. His companion, Lewis, dragged him up into a tree until the crocodiles had gone and it was safe to attempt

Rugged country in the far north. In the distance are the McIlwraith Ranges, location of Cranger's Reef.
PHOTO TOM BUDDEN.

158

Map 38
Some of the many known gold deposits on the Cape York Peninsula.

Rusting machinery at the old Wenlock gold mine, Cape York Peninsula.
PHOTO TOM BUDDEN.

Cape York

C o r a l

Cape Grenville

Gulf

Weipa

Claudie ★
River

S e a

Cape Direction

★ Wenlock

O f

★
★ Hayes
Creek
★

Cranger's Reef
located in this area

Carpentaria

★ Coen
★

Cape Melville

★

★

Potallah Creek ★

Cape Flattery

★

Alice River ★

★
Laura

Cooktown

Maytown ★

Lakeland

★

0   50   100   150   200 km

★

Mossman

Mount Molloy ★

the journey back to Coen. But it was too late, and the delirious Mellone died the day after reaching the town. By this time the story had become legend, with many people believing that Cranger's reef did not really exist. But many others were prepared to risk life and limb for the riches that would result from proving the old prospector's story was true.

At the turn of the century, another expedition set out, this time from Cooktown, led by an Aboriginal guide who claimed to know the exact location of the reef. But in the depths of the ranges the guide disappeared, leaving the party to find their own way out as best they could. Being experienced bushmen, they reached safety, but without finding Cranger's gold and after losing one of their number from snakebite. By now the search for the reef had claimed many lives, but still treasure seekers plunged into the jungle, seduced by the lure of the legendary reef.

One evening in 1929, a stockman, Arthur Ambrust, was relaxing at his camp when an Aborigine approached him and offered a gold nugget in exchange for tobacco. Ambrust could not believe his luck. Whether or not the nugget was from Cranger's mystery reef was not important, the fact that it was gold was sufficient. Ambrust provided the tobacco and offered more if the Aborigine would show him where he had found the gold.

Ambrust followed the man through the jungle and past the swamp that killed Mellone, until, in a clearing, they came upon an old campsite with a miner's rusting pick beside it. After 57 years, Cranger's secret spot had been discovered, for beside the remains of the camp was the fabled reef, with gold-streaked quartz scattered all around.

Cranger's lost reef was subsequently worked, to return a rich treasure of gold. But gold rarely lies in such isolated outcrops and the chances are that somewhere near Cranger's reef is another, perhaps as rich, perhaps even richer.

BEACH TREASURE

A treasure tale which persists and which has all the hallmarks of being authentic, is the story of the beach treasure at Ballina. It is a relatively recent story, dating back only to 1914. Australia was still a foundling nation, having established the Constitution only fourteen years before, and still very much attached to the apron strings of Mother England. When war was declared between Britain and Germany, it was a foregone conclusion that Australians would be totally committed to fight alongside the British armies in every war zone where they were required.

However, this was not quite as simple as it appeared, for already a large percentage of German migrants had made Australia their homeland and were, to all intents and purposes, Australian. Despite their affection for their adopted land, these Germans could not be expected to fight against their own kin. Nor could they be allowed to roam around at large. The only solution had to be internment for all Germans in Australia for the duration of the war.

Some German residents, particularly those in the capital cities, had achieved much in their new land, and a number had amassed sizeable fortunes. Fearing that these would be confiscated when they were interned, they took them to the German Consul in Sydney for transhipment to Germany. Just before war was officially declared, a chest reputed

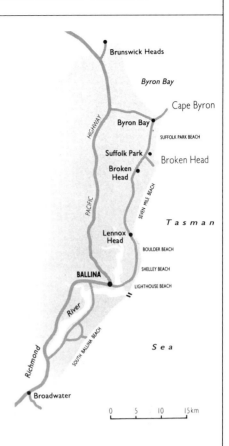

Map 39
The beaches around Ballina.

The shallowing waters off the northern New South Wales beaches cause many problems for boat owners. The seaward progression of the beaches is the most likely reason why Ballina's treasure has not yet been found.

to contain 2500 sovereigns and quantities of personal jewellery was placed aboard the streamer *Sedlitz*, before her departure from Sydney.

The captain of the *Sedlitz* was nervous about his valuable cargo. Fearing that his ship might be captured during the voyage and the precious cargo lost, he decided to bury the chest on a beach on the north coast of New South Wales. It would, he felt, be safe there until the end of the war, when it could be recovered. The beach he chose is thought to be near Ballina, but exactly which beach and where the treasure was buried may never be known, for the captain was killed during the war and apparently the secret died with him.

Locals in the Ballina area who knew the story of the beach treasure searched many of the nearby beaches without success. Since then, many treasure hunters from other parts of Australia have carried out unsuccessful expeditions along the beaches and headlands in the area. In 1938, two Germans who claimed to have knowledge of the location of the treasure arrived at Ballina and began a hunt, but they had no more luck than anyone else. The story of the buried treasure became a legend.

Although no documentation has ever been found indicating either its existence or its location, the story of the beach treasure has a ring of authenticity about it. The most likely source of useful information would be the log of the steamer *Sedlitz* but she was either sunk in the war or broken up shortly afterwards, and no trace of her log has been found. Records in the German Consulate can throw no light on the mystery, nor have any of the *Sedlitz*'s crew come forward with information or, for that matter, endeavoured to find the treasure themselves.

The beaches along the coast near Ballina extend for many kilometres, and are moving; some to seaward as the sand builds up, and some inshore as the beach is eroded away. The site of the treasure may not now be on the present beach, but inshore in the dune infrastructure, or beneath the lines of surf breakers that constantly pound the beach. With modern sensitive metal detectors, the prospect of finding the cache is now much improved. Perhaps a new expedition, armed with the latest in metal detecting equipment, will one day find the treasure which, at today's inflated prices, would be well worth finding.

Map 40
The two major archipelagos of the north-west coast. One of the islands in this area could be the legendary Silver Island.

ISLANDS OF SILVER

The north-west region of Western Australia is one of the richest areas in the continent in terms of natural resources. Huge deposits of iron ore such as that at Mount Newman and Mount Tom Price have brought this distant corner of Australia into world prominence and created new industries and towns where only spinifex and desert grass thrived before. Gold and other precious metals and, in recent years, diamonds, have all added to the enormous wealth of the region. And since this is only the known wealth, it is reasonable to assume that there are many, and perhaps greater, treasures still to be found beneath the brown, dusty soil.

Off the coast, the brilliant aquamarine of the Indian Ocean is peppered with literally hundreds of reefs and islets, some of them the peaks of mountain ranges which were joined to the mainland before the rising seas after the last Ice Age cut them off. Like the mainland, these islands are rich in mineral deposits, some already supporting commercial mining activities, some awaiting their turn to feed the constant chain of ore-carrying ships that ply past them.

Cockatoo Island and Koolan Island are but two where rich mineral lodes have created not only huge mining industries but also major port facilities to export the ore. The Buccaneer Archipelago, in which these two islands lie, and the adjacent Bonaparte Archipelago, stretch across the sea like a chain of scattered gems, any one of which might contain a king's ransom in mineral deposits. Gold, silver and gemstones may well be concealed among the ironstone that forms most of these islands, but there are so many — and some so tiny they do not even appear on maritime charts — that close exploration of every one is not a feasible proposition. Any one of the numerous cruising yachts and small craft that weave their way through the island chains could one day make a lucky strike, landing on a shore that contains a fortune in gold or silver, perhaps even diamonds.

The Macassan fishermen of Australia's earlier history, who sailed these coasts in search of the lucrative trepang, or beche-de-mer, brought back stories of riches on the shores of the unknown south land. Most of those legends were probably the result of relaxed story-telling around camp fires on the tropical beaches. Others may have been acquired from encounters with Aborigines on the mainland.

One such story concerns a trepang fisherman from what is now Indonesia. While searching for his prey among the islands of the north-west, his boat was struck by a cyclone and blown well off course, finally grounding on an uninhabited island. To make the boat more stable, the crew loaded aboard some large rocks which lay along the shore. On their return to port they were astonished to find that the rocks contained a very high grade silver ore. After selling the ore at a good price, they returned to find the island again, but to no avail. The cyclone had set them

Luggers tied up at Broome. It was in boats such as these that the legend of the treasure islands originated.
PHOTO JOHN AND DEDE DEEGAN.

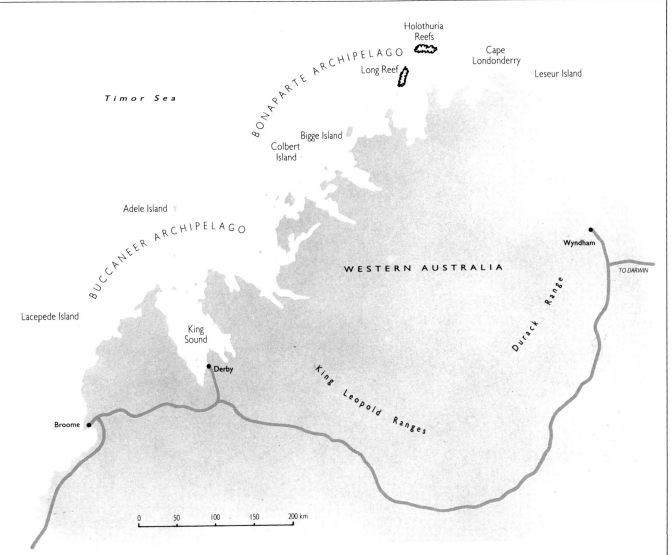

so far off course that they could not retrace their track.

The story of the silver treasure may well have passed into history had it not been for an incident soon after the turn of the century. A police party from Broome, investigating a murder aboard a lugger in King Sound, came upon a large fuel drum full of silver. Since the owner had been murdered by Aborigines after landing on the shores of King Sound, the secret of the silver died with him. But the coincidence between this occurrence and the legend of the silver island in the offshore archipelagos is too great to be easily dismissed. King Sound lies inshore from the islands of the Buccaneer Archipelago.

Once again, the legend of the silver island may have drifted into obscurity, for although numerous locals from Broome made extensive searches of the islands, no silver lode was found. But then, in 1920, two German adventurers who had hired a pearling lugger from Broome returned with a story of having been struck by a storm and forced to put ashore on an island in the outer archipelagos. In a remarkable duplication of the earlier story, they had loaded the lugger with large rocks for ballast before returning to Broome.

For some time the rocks lay unnoticed in the hold of the lugger, for this was a common method of ballasting a ship when cyclones were around. Some years later, when the boat was sold, the new owner removed the

rocks, to find that they were rich with silver. A new and frenzied search of the islands began, but to this day no reports of huge silver ore deposits have come to light. Somewhere out there, among the hundreds of reefs and islands, is an island that contains a fortune in silver. Some day some adventurous treasure hunter will find it. Not only will he be rich beyond his dreams, but the silver island will at last have been proved to be fact not legend.

Wreckage of a World War II bomber indicates the isolation of these northern islands. Such a relic near more settled areas would long since have been torn to pieces by vandals and souvenir hunters.
PHOTO JOHN AND DEDE DEEGAN.

THE MAHOGANY SHIP

One aspect of Australia's history that has engrossed historians for almost two centuries is the possibility that the east coast of the continent was discovered by Europeans before the arrival of Captain Cook in 1770. At that time, much of the south, west and north coasts had been discovered by Dutch and Spanish navigators. It seems reasonable to believe that other parts of the continent, including the east coast, could have been visited by a ship or ships from one of these maritime nations.

The Spanish, in particular, could have been searching for a new trade route between the Pacific and Indian oceans, since their trade with Mexico, South America and the Philippines had to run the gauntlet of pirates and hostile navies that were prominent in the waters of the East Indies. The Portuguese may also have been trying to extend the discoveries of Magellan by seeking the possibility of a route around the south of Australia, similar to the strait he found at the southern tip of South America.

Three factors point to the possibility of Spanish or Portuguese contact with the south coast of Australia. Firstly, in a tiny, sheltered inlet called Bitangabee, on the south coast of New South Wales, ruins of a settle-

ment have been found which, experts agree, was built before Cook sailed along the coast. The indications are that they are of Spanish or Portuguese origin. Secondly, in a limestone quarry in Port Phillip Bay, keys were unearthed which have also been identified as Spanish or Portuguese. The level at which they were buried is believed to coincide with the level of the beach in the early sixteenth century.

About this time, the Portuguese explorer Cristovao de Mendonca was thought to be in the southern seas. Equipped with a flotilla of three caravels, small but hardy ships often built of mahogany, Mendonca was on a roving commission which could well have included exploration of the unknown continent referred to by navigators as the Great South Land. If he were connected with the Bitangabee settlement — perhaps a base camp — and the keys in the limestone quarry at Port Phillip, then it would seem that Mendonca passed through the Bass Strait.

The third and vital factor indicating the possibility of Portuguese navigators on the south coast of Australia is the legendary "Mahogany Ship", believed to be a Portuguese caravel, and lying beneath the sand dunes near Warrnambool on the Victorian coast.

The first sightings of the Mahogany Ship are not documented and therefore somewhat vague. In his book *The Secret Discovery of Australia*, Kenneth Gordon McIntyre lists the first discovery of the wreck in 1836, two years after the colonisation of Victoria. Three sealers out of Launceston, working the coastline near the present site of Warrnambool, were wrecked on the bar at the entrance to the Hopkins River and one was drowned. Since there was no settlement on this part of the coast, the two survivors headed for the Moyne River, now Port Fairy, about 30 kilometres to the west, where a sealers' depot had been established.

While walking along the beach, they were surprised to come upon a wreck lying between the first and second rows of sand dunes. Although wrecks were not uncommon along this poorly charted coast, this wreck was no sealer or whaler, but of an ancient design and built of an unusual timber. They reported their find to a Captain Mills, one of the earliest residents at Port Fairy, who, together with his brother, examined it and concluded that it was an ancient ship and built of mahogany. Another sea captain, James Mason, examined it in 1846 and established that it

Map 41
Last known site of the Mahogany Ship (taken from an old map).

The beach claims yet another victim. A stranded whale lies rotting on almost the exact spot where the Mahogany Ship was last seen.
PHOTO PAUL HANKE.

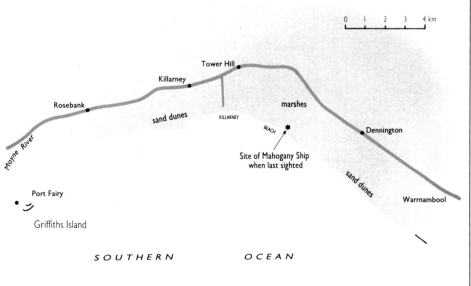

165

was most likely of Portuguese origin, about 100 tonnes burden and built of either mahogany or cedar. It was from these reports that the wreck gained its name Mahogany Ship.

All that remained of the ship was its hull; the decks and spars had disappeared, probably because they were made of softer wood. The hull was of a wood that was as hard as iron, as Captain Mills commented when he tried to cut off a sample with his knife. Captain Mason was of the opinion that although old, the ship was not a galleon since it did not conform with the standard design for those ships, lacking the high prow and deep sheer which were a feature of the Spanish galleons.

The description from both captains, the position and condition of the wreck, all lend themselves to the theory of a Portuguese caravel, wrecked on the beach some hundreds of years earlier. Mendonca had commanded a flotilla of caravels, and old Portuguese maps, which were compiled partly from his records, follow the coast of Victoria to the approximate location of Warrnambool and then stop. The inference must be that the Mahogany Ship was one of Mendonca's flotilla of three caravels, and that after its loss, Mendonca turned back from his exploration of this coastline.

In all, 27 different people claim to have sighted the Mahogany Ship before it was lost from view. A huge storm one night in 1880 changed the contours of the sandhills along the beach between Port Fairy and Warrnambool. When it died down there was no trace of the wreck, although some ironwork and planking has been found since. Numerous expeditions, including one by the Monash University Archaeological Society, have failed to find the wreck, although a bronze spike and iron latch were found at the site. With sensitive modern metal-detecting equipment it is not outside the bounds of possibility that the metal in the ship's hull could provide the final clue as to her resting place, and lead to her eventual recovery.

Because of the authentic sightings and the volumes of information about the ship recorded in past newspapers, books and magazines, the Mahogany Ship ranks as one of the most exciting of all hidden treasures in Australia. A treasure that, when it is found, could change the face of early Australian history.

LASSETER'S LOST RICHES

A more inhospitable spot than Surveyor Generals Corner would be hard to find anywhere in the world. It is dry, dusty, bleak and with no vegetation other than desert scrub. To the south is the Great Victoria Desert, to the north the Great Sandy Desert, and to the west the notorious Gibson Desert. Ayers Rock lies only 150 kilometres to the north-east. This is the red heart of Australia.

Surveyor Generals Corner is the point where the boundaries of Western Australia, South Australia and the Northern Territory meet. Apart from a few mineral prospectors and the odd adventurer, few white men wander into this country. It is mostly arid and supports only life forms such as the familiar Horned Devil, a few monitors and other creatures that have adapted to the dry, dusty environment. Desert grasses and mulga scrub cover much of the landscape, although in the

Spinifex, desert scrub and rugged red mountains. Lasseter prospected in this area near the MacDonnell Ranges before discovering his legendary gold reef.

PHOTO ROBYN HILL.

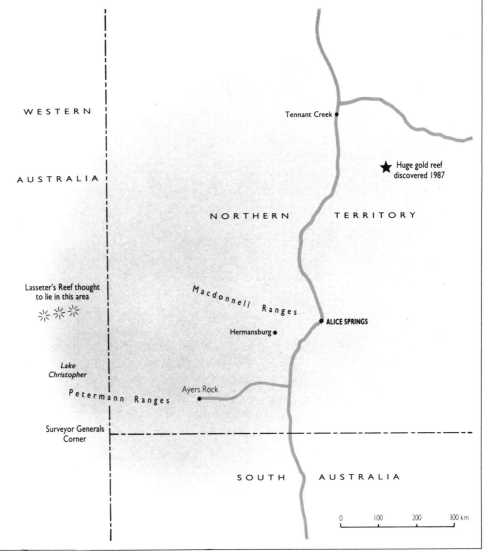

WESTERN

AUSTRALIA

Tennant Creek •

★ Huge gold reef discovered 1987

NORTHERN   TERRITORY

Lasseter's Reef thought to lie in this area

Macdonnell Ranges

ALICE SPRINGS •

Hermansburg •

Lake Christopher

Petermann Ranges   Ayers Rock •

Surveyor Generals Corner

SOUTH   AUSTRALIA

0   100   200   300 km

Map 42
The area of Central Australia where Lassester searched for his fabled reef.

undulating hills to the east, ghost gums indicate the presence of a water table not far below the surface.

To the north and east of this barren corner are a series of mountain ranges. Pushing upwards through the red soil, they provide a stark contrast to the dreary, lowland desertscape. Massive red escarpments of iron-stained quartzite rear dramatically in ragged rows as though the enormous geological upheaval that folded and broke the earth's crust and pushed them skywards, occurred only recently. So spectacularly beautiful are these mountains that many areas have been preserved for all time as national parks or conservation areas.

But the beauty of the red rocks lies not only in their outward appearance. Hard siliceous rock such as quartzite is favoured by prospectors for its association with gold deposits. And the lure of gold has brought many treasure seekers to these mountains. The MacDonnell Ranges, which sweep to the east past Alice Springs, have long been the venue for prospectors looking for minerals and gems. Farther to the south, the Petermann Range, not as visually spectacular as the MacDonnells, but with equal promise of gold, was the scene around which one of Australia's greatest mysteries was played out — the mystery of Lasseter's Reef.

Probably no story of hidden treasure has caught the imagination of the public as much as Lasseter's Reef. Supposedly first discovered at the turn of the century, the mystery of the reef has lost none of its fascination in almost 90 years of story-telling and treasure-seeking. Despite innumerable expeditions, some small, some with huge financial backing, Lasseter's Reef has never been found. Yet the enthusiasm never wanes, and almost every year a new expedition is planned to locate this elusive reef and lay to rest for all time the old prospector's mystery.

Harold Bell Lasseter was an Australian who became a naturalised American. Little of his past is known either as an American, or when he returned to Australia. He had obtained full qualifications as a surveyor while overseas and came back to try his hand at prospecting in the far west of Australia. For the most part he prospected alone and kept to himself, revealing little about what he had found on his expeditions. Indeed, so reluctant was he to discuss his activities, that for almost 30 years Lasseter kept secret what he claimed to be the biggest gold strike in Australia's history.

The story began in 1900, to the west of the magnificent MacDonnell Ranges, where Lasseter had been fossicking for rubies. He set off to cross to the Western Australian coast but became lost and wandered westwards through the desolate area to the north of Surveyor Generals Corner. Emaciated and dying of thirst, he was found and brought back to civilisation by an Afghan camel driver. Later he confided to a fellow surveyor that during his ordeal he had stumbled across a massive reef of gold, some 16 kilometres long.

It was not until 1930 that Lasseter announced the find publicly, and the long delay, plus the immense size of the reef, damaged his credibility. The claim was ridiculed as the hallucinations of an old prospector who had suffered too much exposure from the sun. However, there was one factor which gave credence to the story: when Lasseter was brought in from the desert, he clutched in his hand a bag of rock samples richly embedded with gold!

It was probably this, more than anything, that gripped the public imagination. If the prospector had a bagful of gold, he must have struck a reef. And whether it was 16 kilometres long or 16 metres long, it was gold. A wave of gold fever swept across the country and expeditions of all sizes set out to locate Lasseter's mystery reef, which he claimed lay in an area somewhere between the Petermann and Ehrenberg ranges.

Most significant of all these expeditions was conducted in 1930, when Lasseter himself was engaged to act as guide. A syndicate of Sydney speculators floated a company called the Central Australian Gold Exploration Company Ltd to search for the fabled reef. It was a well-organised, well-funded expedition, equipped with a suitable vehicle, every facility that might be required, and even an aircraft for aerial reconnaissance. The party was led by an experienced explorer, Fred Blakeney, and included Lasseter, an engineer, another explorer and a pilot. At Alice Springs, two more men, one an Aborigine who knew the country well, were added to the group.

A camel expedition sets off across the desert. Lasseter used camels for his last fatal push into the unknown.
MITCHELL LIBRARY, STATE LIBRARY OF NSW.

But problems struck from the very beginning. On the way to the search area, the plane crashed and the injured pilot had to be rushed back to Alice Springs. The rest of the party pressed on, but after several days of slow progress, Lasseter decided they were looking in the wrong spot. The reef, he estimated, was some 240 kilometres to the south. As the party moved south, their vehicle became bogged in a series of sand ridges and could not be moved. However, in the meantime the company had organised a replacement aircraft, so the search was to continue by air. Flying low over some distinctive landmarks, Lasseter claimed to recognise the area as the one in which the reef lay. But the aircraft was at the limit of its range, so the search was called off while the plane returned to Adelaide to have long-range fuel tanks fitted.

Now fate began to play its hand — the expedition ran short of money. Most of the group were ordered to return to Sydney, leaving Lasseter to make another attempt to find the reef, this time with a camel team and

its handler, Paul Johns. The refurbished aircraft returned, but crashed in the desert and had to be abandoned.

Lasseter and Johns struggled on with the camels, but the country became too difficult even for these hardy animals. When they finally called a halt, Lasseter claimed they were within 32 kilometres of the reef. It was decided that he would go on alone, while Johns carried the supplies back to their base camp at Lake Christopher. When Lasseter had found and pegged the reef, he would return and rendezvous with Johns at the lake.

Lasseter was never seen alive again. A search party found his notes and his diary, buried beneath the ashes of his camp fires, and from these the final segment of his story was pieced together. It appeared that shortly after leaving Johns, Lasseter found and pegged the mystery reef. He pointed out in his notes, however, that following parties would not find the pegging as the reef was located within a sacred site and the Aborigines would remove the pegs as soon as he left the scene.

Lasseter had attempted to return to civilisation but on the way his camels bolted, leaving him alone in the desert without food or water. He was befriended by an Aboriginal group with whom he lived for some months. But the tough life of the desert nomads took its toll, and when the search party did not arrive, Lasseter, by now sick and weak, decided on a last desperate gamble. Alone, on foot, he headed back towards Alice Springs but collapsed at a spot in the Petermann Ranges known as Winter's Glen. He died alone and in agony, and the remains of his body were not found until some weeks later, when a search party found and buried them in a rough grave.

Since Lasseter's death, the interest in his story has never waned, and in fact has become more fascinating with the passage of time. Research has raised many questions about the fabled reef and thrown new light on some facets of the story. But all expeditions that have followed in the prospector's footsteps have come back empty-handed. One such expedi-

Typical of the arid country in which Lasseter searched and died. Not even rough desert roads provided guidance in those days.
PHOTO ROBYN HILL.

A memorial to Lasseter's last journey erected by the Docker River Social Club Inc for Mr R. Lasseter. The legend reads: Lewis Harold Bell Lasseter sheltered in this cave for approximately 25 days during January 1931. He was stranded without food after his camels bolted at a point 15 km east of here. Although weak from starvation he set out about 25th January to walk the 140 km to Mount Olga, hoping to meet up with his relief party. Carrying 1.7 litres of water and assisted by a friendly Aboriginal family he reached Irving Creek, in the Pottoyu Hills, a distance of 55 km, where he died about 28th January 1931.

PHOTO TOM BUDDEN.

tion, in 1931, included the Western Australian Government geologist, who stated categorically that no gold would be found in the region. Yet geologists have been wrong before and many rich gold strikes have been made where "no gold would be found".

One expedition in 1931 failed to find gold, but found instead a substantial outcrop of mica — fool's gold. The deposit was in an Aboriginal sacred site in the general area of the Petermann Range. However, Lasseter was a qualified surveyor and experienced prospector and it seems unlikely he would have been deluded by the yellow-coloured mica.

The two most convincing aspects of Lasseter's claim must lie firstly in the samples of rock he brought back, and secondly in his own unwavering conviction that the reef existed. If his story was fiction, then where did the gold-laden rock samples come from? To produce gold samples, he must have found gold somewhere, and there is no shortage of evidence that the samples were genuine.

And similarly, if the whole thing were a hoax, would Lasseter have staked his own life on it? He was in the forefront of the main expedition which searched for the reef in 1930, knowing, as he did, that the outcome could easily be death. When everything had failed, and even Johns had turned back with the camels, Lasseter pressed on alone, intent on proving nothing to anyone except himself. This was surely the action of a man with the courage of his conviction.

In 1987 Ray and Maria Hall, a prospecting family with a mining lease some 200 kilometres south-east of Tennant Creek, struck a gold reef that immediately revived the Lasseter legend. The new reef has barely been worked, but already is revealing one of the richest finds in Australia's gold history. More than 2 metres wide and with gold content as high as 2835 grams (100 ounces) to the tonne, the full extent of the reef is not yet known, but these characteristics alone revive memories of the desperate prospector forfeiting his life to rediscover the fabled reef he claimed to have found.

The Hall reef is nowhere near the area in which Lasseter placed his reef, yet when the search was under way in 1930, he at one stage amended his original figures and put the location of the reef some 240 kilometres to the south. If he had amended the figures 240 kilometres to the *north*, the search would have been closing towards the area in which the 1987 strike was made! And Lasseter is known to have once prospected in the area.

Could the mystery of Lasseter's reef be simply a matter of confused bearings? Was the old prospector's memory addled by the privations he had suffered, his brain affected by the sun? Probably the answer will never be known. But the Hall strike in 1987 proves that some ninety years after Lasseter staggered in from the desert with his tale of a fantastic reef, there is still a possibility that he was right. To this day, rich gold reefs are waiting to be found in the arid land where Lasseter lived and died searching for the one he found and lost.

TREASURE ISLANDS OF TORRES STRAIT

For 164 years before James Cook sailed his ship *Endeavour* around the northern tip of Australia, Spanish ships had negotiated a westward passage through the Torres Strait. Discovered in 1606 by Luis Vaez de Torres, the Strait provided a short cut for merchant ships returning to Europe from the Pacific, and for this reason was kept a secret by the Spanish in order to steal a march on their trading rivals.

How many Spanish ships sailed through the Strait will never be known, for records were not kept, but it is certain that although many made a successful passage, many came to grief. The tortuous passages through poorly charted coral reefs, and the extreme tidal phenomenon that exists, particularly at the eastern end, claimed many a fine vessel with its cargo of riches from the Spanish trading ports in Mexico, Chile, or the Philippines. Pirates and buccaneers, preying on the treasure ships, also knew of the passage, which could provide them with an escape route if they were tracked down and hounded by a naval ship.

But honest traders and pirates are alike in the eyes of the sea, and a pirate ship can be smashed to pieces on a reef as quickly and as devastatingly as can a giant treasure-carrying galleon. The ferocious jaws of the Torres Strait waited indiscriminately for any ship that tried to run the gauntlet of its tides and reefs. With poor charts and no navigational aids, the victims were many and the survivors few.

How many made a successful transit and how many fell victim to the treacherous waters only the reefs and islands know, but the existence of those ships is not in doubt, for even to this day an occasional relic, linking the present to that distant past, comes to light on the shores of islands in the Torres Strait.

Boot Reef

In the early days of European settlement on Cape York, relics were often found along the white beaches and in the many bays that line the Torres Strait. Spanish coins and artefacts, washed ashore from a sunken treasure ship or a successful pirate raid, gave credence to the stories handed down by the Islanders of shipwrecks and murder and treasure-troves in the reef-studded waters. Few treasure hunters took the trouble to follow these stories through, and yet the evidence seemed credible enough.

Indeed, if evidence were needed, the find, in 1890, of Spanish silver and gold coins in a coral lagoon at the eastern end of the Torres Strait

Submerged coral reefs at the tip of the Great Barrier Reef add to the dangers in the eastern approaches to the Torres Strait.

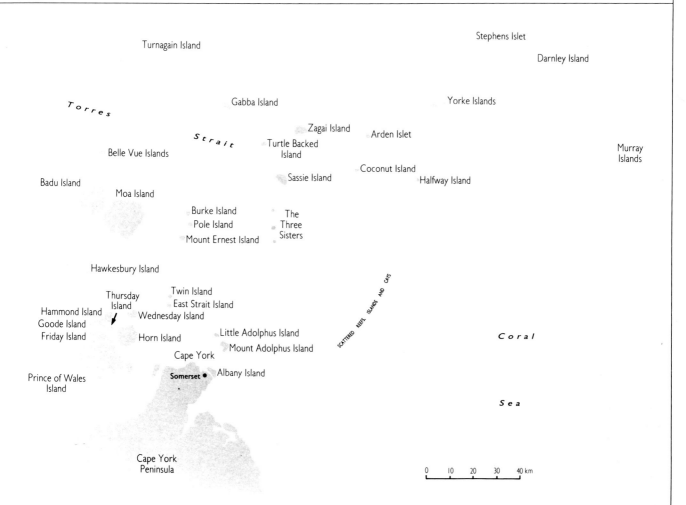

Map 43
Torres Strait

would remove all doubt as to the presence of Spanish treasure in the region. A similar treasure was unearthed from another reef some years later. In both cases the finds were made by beche-de-mer fishermen employed by the pioneering Jardine family.

The Jardines were instrumental in opening up the Cape York Peninsula to white settlers. After epic overland journeys to bring stock up to the new pastures, they established their own property at Somerset, close to Cape York itself. Apart from their rural interests, the Jardines operated a fleet of luggers in the Torres Strait waters, collecting beche-de-mer for the lucrative Chinese market. It was one of the Jardine fishermen, walking along a coral reef at low tide, who discovered the cache of silver and gold coins lying in a cleft among the coral. When retrieved and cleaned, they proved to be of Spanish origin, mostly silver dollars with some gold coin.

A few years later, another of the Jardine luggers, this time under the command of a Captain Samuel Roe, was working the same area when it was forced to run for shelter before an approaching cyclone. Since there are no havens of consequence in the area, Roe took his boat into a lagoon inside Boot Reef, in the eastern mouth of the Torres Strait. The boat rode out the cyclone safely, but when the crew attempted to take her out of the lagoon, they found there was insufficient water. In their haste to find shelter, they must have crossed the reef on an exceptionally high tide. Now that the tide had dropped, the boat was trapped inside the lagoon.

Roe, experienced in the ways of these waters, put divers over the side to hack a passage through the coral with picks and axes. The channel was half completed when one of the divers surfaced in great excitement. Throwing a handful of coral and rock onto the deck he pointed out a bright glitter of metal among the debris. In breaking up the coral reef he had unearthed a mass of silver coins, which, when cleaned, again proved to be Spanish dollars.

When the passage was finally cleared, Roe sailed his lugger back into port with the treasure piled up on deck. Frank Jardine confirmed that the coins were genuine treasure-trove and divided them fairly between the crew. His own share was melted down in Sydney and made into a fine dinner service for his table at Somerset.

The isolated homestead at Somerset was a unique feature in Australian early life. Situated on the shores of Albany Passage, close to Cape York, it was passed by most of the mail and passenger ships heading north around Cape York to Europe and India. A stopover at Somerset provided a welcome change for passengers, many of whom were rowed ashore to be entertained by the Jardines. As the hospitality of the isolated outpost became known, more and more ships called at Somerset, and the Jardines' guest list became a who's who of Australian public life and included most of the dignitaries who travelled to and from the colonies.

It would be interesting to know if, in the course of his dinner parties, Frank Jardine enlightened his guests on the origin of the dinner service. Or whether he left them blissfully unaware that the cutlery they were using may have once been part of a king's ransom, or, on the other hand, part of a pirate's booty.

Like so many of the treasures and relics discovered in the north, there has been no attempt to follow-up the discovery of the "Jardine Dollars", as they have become known. Captain Roe is reputed to have taken his lugger back to Boot Reef in an attempt to find more treasure, but the story is not well documented and apparently all hands on the ship, including Roe, were lost under mysterious circumstances.

Although Spanish dollars were the universal currency of the period, and the treasure could have originated from a ship of any nationality, it seems unlikely that any but a Spanish ship would be carrying such large quantities. Also, the Spanish were very tight-lipped about the discovery of the Strait by Torres, and it is equally unlikely that any other European ships were in the area. Not only was it an uncharted and remote corner of the isolated southern oceans, but no other nation had colonies or trading partners in the area, or in an area of the Pacific Ocean which would make a passage through the Torres Strait useful.

The South American, Central American and Philippine regions were all under Spanish influence, which made the passage an important maritime sea lane for Spanish ships, just as the ships of the Dutch East India Company used the favourable winds along the coast of Western Australia to reach their colonies in the East Indies. And just as the Dutch ships carried specie to finance their settlements and fund their trading activities, so the Spanish carried gold and silver coin, jewellery and other valuables in the ships which sailed through Torres Strait. Pirates, such as Benito Bonito (see Chapter 8), who plundered the treasure ships and sacked the cities of the richer colonies, were also active in these waters. In later years, when the useful Strait became known to other mariners, American ships trading with China also used it.

Murray Island

Small wonder, then, that Torres Strait, with its scattered islands and reefs, many of them uninhabited, is a potential treasure-hunter's paradise. Legends handed down through generations of Islanders enhance the treasure-seeker's dream. On a reef near Murray Island a ship with a silver keel is supposed to lie awaiting an intrepid treasure seeker. The silver keel is an unlikely prospect, since no ship in recorded history has been so sumptuously equipped. But the distortion of retelling the legend over centuries could mean that a ship with the bottom of her hull full of silver ingots or coins is lying in the deep water off Murray Island; a fine treasure for an enthusiastic treasure hunter.

Murray Island is one of a gaggle of islands known as the Murray Group, situated at the eastern entrance to Torres Strait, near Boot Reef, described earlier in connection with the Jardine treasure, and not far from Stephens Island, which also is connected with treasure-trove. These islands and reefs, guarding the entrance of the Torres Strait against the onslaught of the Pacific Ocean, create a virtual death trap for approaching vessels. Even to this day, the eastern approach channels that wind past these offshore coral traps are considered a major navigational hazard. In the days when bulky, cumbersome trading ships, deep-laden and hard to manoeuvre, made their run into the Strait, many fell prey to the waiting dangers. Like King Island, in Bass Strait, these island groups became the graveyard of fine ships, their crews and their cargoes.

A legend handed down by the people of the area, and one which has been given credence by discoveries of relics in recent years, is that of the treasure on Murray Island. This time the connection is not with an early Spanish ship, but a vessel of an unknown nationality, but presumably European, which came to grief on one of the reefs somewhere in the eastern approaches to the Torres Strait. The survivors, clinging to a raft which also carried several chests of treasure from the ship, landed on Murray Island and buried the treasure. Some days later they were killed and eaten by cannibals, and the buried treasure passed into legend.

As is so often the case, the legend was revived many years later when coins were found both on Murray and an adjacent island in the Murray Group. In relatively recent years, an hour glass and a silver dollar were found on Murray Island, which seems to confirm the story of a wreck in the area. Once again the remainder of the treasure has never come to light and may still be buried where the ill-fated survivors placed it, awaiting discovery by a keen treasure hunter.

Green Island, a Barrier Reef island south of Torres Strait, is a typical example of a coral cay surrounded by a reef.

Stephens Island

Stephens Island has intrigued treasure hunters since 1934, when a trochus shell fisherman, exploring a beach on the island, discovered a carved stone idol of the native god Zogo. Digging in the sand around it, the fisherman came across a quantity of jewellery including a number of rubies which appeared to have come from an ornate necklace. The discovery was hailed as conclusive evidence that the legend of Stephens Island was not in fact a legend, but a true story, and that the remainder of the treasure must still be somewhere on the island.

Stephens Island is a tiny coral cay on the northern side of Torres Strait. The details of the story are lost in time, but the legend told by the Islanders would appear to date back to the time of the Spanish ships that plied through the Strait en route to Manila, Mexico or South America. A vessel, believed to be Spanish, was caught in a storm and wrecked on a reef near Stephens Island. A few survivors, including the captain and his wife, made it to the shore where they huddled in fear as hostile Islanders surrounded them. Their fears were well-founded and before many hours had passed, the miserable survivors were attacked and slaughtered; only the captain's wife was spared.

Among the jewellery which she had saved from the wreck was a magnificent necklace studded with a number of rubies, including one very large ruby which formed the centrepiece. Whether or not it was this necklace that saved her from the massacre, no one will ever know, but there was no question that the natives coveted the beautiful red stone. In the course of a tribal ceremony later that night, they removed the necklace and draped it over a stone idol of their god Zogo. Entranced by the fire of the big ruby, the natives drummed and danced themselves to a frenzy as they dedicated the glowing red "eye" to Zogo. The woman took advantage of the distraction and slipped away. She was never seen again, and since the island was searched many times, she presumably ran into the sea and was drowned.

In later years when either the islanders tired of Zogo, or he was replaced by a god brought by the missionaries, the idol and the jewellery were buried at the beach. The final relic of the gruesome ordeal was gone and the tale passed into legend. But only until, more than a century later, a fisherman unearthed Zogo, complete with his rubies, and brought it to life again. Now the legend became fact, to tantalise treasure seekers with the lure of a real treasure-island cache.

The captain's wife almost certainly had jewellery other than the necklace, and it is more than likely the captain retrieved money and valuables from the ship before she was abandoned. The fisherman recovered only the rubies, so somewhere on Stephens Island a fortune in Spanish coin and jewellery may be awaiting an intrepid treasure hunter.

Prince of Wales Island

To the Torres Strait Islanders, the stories of ships lost in the tortuous waters that surround their islands have been part of their heritage for centuries. The tales have been enhanced from time to time by finds of relics or coins, either on the reefs where they dive for fish or on the beaches of the many islands. Indeed, so common were these finds in past generations that inducing the Islanders to reveal family "heirlooms" and personal treasures handed down from generation to generation, would probably bring to light a small fortune in Spanish coins and relics. Unaware of their value, the Islanders of the past would have kept the relics as personal bric-a-brac or used them as ornamentation for special occasions.

Even within living memory, numerous finds of Spanish dollar coins have been made on the beaches of the Strait islands. Particularly is this the case with Prince of Wales Island, where silver coins were found not infrequently on the beaches that face Thursday Island, indicating the possibility of a Spanish wreck somewhere in the channel between the two. Another of the Islanders' legends relates to a treasure buried somewhere on the island itself, perhaps the main part of the ship's treasure, saved by her crew when they abandoned her.

More tangible evidence of Spanish intrusion in the area came in 1926, when a skeleton was discovered in a cave on Prince of Wales Island. The skeleton had a sword beside it, later identified as of Spanish design, and a gold goblet nearby. There is, of course, nothing in this discovery alone

to suggest a shipwreck in the locality, for the man could have been a castaway or survivor who had sailed a boat many kilometres from a distant shipwreck. But when taken in conjunction with the coins found on the beach, it points fairly strongly towards the possibility of a Spanish wreck near the island.

The Torres Strait, for so long a secret passage known only to the Spanish, must be credited with the greatest potential for successful treasure hunting of all Australia's treasure grounds. Isolated, and in waters with treacherous tidal flows and cyclones in summer, the islands and reefs of this waterway are not popular areas for modern treasure seekers. Yet danger is often the greatest incentive to intrepid adventurers, and the possibility of success would seem to be greater here than anywhere else. In the dry season, and with careful attention to winds and tides, there is no reason why a skilful navigator could not weave his vessel safely through the maze of reefs and islands to establish a treasure-hunting operation.

It is one of Australia's last frontiers in terms of treasure-trove, and this in itself should stir the blood of any true treasure seeker. Since the area is virtually untouched and largely uninhabited, who knows what treasures might lie beneath the crystal clear waters. Treasures that might not only provide a good reward for the adventurers who seek them, but also shed new light on the pre-Cook history of Australia's north.

A lugger makes through the Prince of Wales Channel. A number of Spanish wrecks are thought to lie in this passage.
PHOTO JOHN AND DEDE DEEGAN.

Right: Murray Island, near the eastern entrance to Torres Strait, is rumoured to be the site of a buried treasure.
PHOTO WELDON TRANNIES.

CHAPTER TEN

TODAY'S
TRASH
TOMORROW'S
TREASURE

Every now and then someone makes newspaper headlines by cleaning out their attic and finding an item that creates a sensation when put up for auction. More often than not it is an old book, a historic artefact, or an item of jewellery that was stowed away by a long-dead relative and forgotten. Such finds are less common in New World countries such as Australia, than in Britain, Europe and other countries where civilised history goes back over many centuries.

Many of today's Australians who were born in Old World countries will carry memories of childhood visits to grandparents who lived in old houses with small windows and low doorways, with dingy rooms dimly lit by candlelight or oil lamps. Going to bed was often a fascinating experience, climbing steep, winding stairs to a room with sloping ceilings, with a huge china washbowl that was filled from an equally huge jug, and that unmentionable thing under the bed. Lying in bed watching the flickering candle conjure up all manner of frightening shadows on the low ceiling, or getting up early and leaning out of the tiny bedroom window to pick ripe pears off the espaliered tree on the wall; these

Keys of different shapes and sizes can make an interesting display.
PHOTO SCOTT CAMERON, COURTESY BORONIA ART GALLERY, MOSMAN.

Opposite: Old tins and boxes carrying once-familiar labels are immensely popular with collectors.
PHOTO SCOTT CAMERON, COURTESY THE COUNTRY TRADER, PADDINGTON.

were all experiences that at the time spelled excitement and to this day remain nostalgic memories.

But most exciting of all was exploring the attic. This was an adventure on a par with visiting King Solomon's Mines or Indiana Jones' Temple of Doom. Dim, claustrophobic and festooned with cobwebs, the attic was at once a frightening, romantic, exciting and stimulating adventure. Older children rustled up extremes of courage in order to climb the ladder and lift the creaking trapdoor, while younger children clung leech-like to siblings' skirts or trouser legs as they were dragged, terrified but willing, into the dark abyss. Entering Grandma's attic was to step into an unknown world.

But once in, with eyes more accustomed to the gloom, the adrenalin slowed and eager anticipation began, for the Devil's Crypt suddenly took on the appearance of an Aladdin's Cave. The headless monsters shrank back into the corners and revealed a potential treasure-trove. Nervousness was replaced with excitement as the lids of old chests and trunks creaked upwards and the hoarded trivia of generations gone by was laid open for exploration. It was an adventure that was always rewarding, for somewhere in the collected paraphernalia of Gran and Grandpa's life was inevitably something of interest, even for kids.

Nowadays those attics are treasure-houses for adults. The memorabilia that was gathered, squirrel-like, by grandparents can have considerable value as collectors' items or even antiques. Because of this, and probably also because modern houses do not have romantic attics in which to store yesterday's trash, finding such an attic is rare indeed. Particularly in Australia with its relatively recent history, and its homes designed to remain cool in summer which rarely use the space immediately under the roof.

However, most human beings are squirrels by nature and even in the most modern homes there are storage areas of some description, cluttered with sea trunks or suitcases or removalists' boxes. And often these contain just as many treasures as Granny's attic. Indeed, many of the treasures originated there and have been handed down and stowed away because they did not suit the decor of the house or the fashions of the time. Town and city homes can have as much potential as rural homes and properties when it comes to hoarded treasures, albeit usually on a smaller scale.

When the old boxes, suitcases and trunks are turned out, the treasures they reveal may find their way into school fêtes, local fairs, garage sales or auctions. Knowing what is trash and what is treasure can prove rewarding, for many an unrecognised treasure surfaces on the tables of a fête or auction. Dealers and collectors haunt these places, always on the lookout for a bargain or, even better, a treasure.

The bookstall, for example, is always popular at the local church or school fête. Most books sell for a fraction of their original value, yet some can be of considerable value to a collector. Old comics, which are often sold off in cheap lots, will come in for close scrutiny from comic collectors, for a single comic, if it is of the right vintage and type, can fetch several hunded dollars in the collectors' market. There is no better source of cast-off comics than school and church fêtes, and although the chances of finding a valuable item at a bargain price are small, half the fun, as with all treasure hunting, is in the search.

Treasure is often in the eye of the beholder and a day spent searching through Granny's attic, the storage room under the roof, or at the local fête, may be rewarded in a number of ways. The discovery of an item that has cash value is always rewarding, but finding treasures of aesthetic value can be equally rewarding if they give pleasure or satisfaction. Like everyone else, treasure hunters' tastes differ and a piece which is passed over by one may delight another, even if, as mentioned, it has no monetary worth. For that matter, just spending the day sifting through a pile of memorabilia can create a sense of satisfaction, even if no tangible treasure comes to light, and many treasure seekers find this sufficient reward for their efforts.

Some treasure hunters are collectors, and therefore specialise in searching for items that can be added to their collection. Some collections can be extremely valuable and provide a worthwhile hedge against inflation. Paintings arc an example of this, as are clocks, gold and vintage cars. But these are the up-market end of treasure hunting and collecting, and for the most part involve expense that can only be justified if the collector is an expert in his field. It is possible to build up a collection without spending large sums of money, but it is still important to have a reasonable knowledge of the items being collected.

It is not possible, in a book such as this, to create instant experts. Years of learning and experience are necessary. But it is possible to provide amateur treasure seekers with a brief background of the major collectables, albeit across a relatively limited range. Knowing a little about the more commonly collected items and their value in the market place will make the hunt more interesting and exciting.

Book collectors are among the most avid treasure-seekers, particularly at fetes and markets. Old volumes such as these would be snapped up very quickly — from left to right, W. L. Buller's *Manual of the Birds of New Zealand*, published in Wellington, New Zealand, in 1882; *Frontier Life: Taranaki, New Zealand* by E. S. Brookes, published in Auckland in 1892; and *Australasia's Story* by H. E. Marshall, published in London.

BOOKS

Books can be among the most exciting treasures to find. There are literally thousands of places where old books may turn up, and finding and collecting them can be great fun. It is a form of treasure hunting that can be enjoyed by the old as well as the young. It requires nothing in the way of specialised equipment and is inexpensive, providing it is kept at a reasonable level. Enthusiastic collectors may pay hundreds, even thousands of dollars for a rare book, but the average treasure hunter, whose thrill comes from the search as much as the possession, can keep cost down to a minimum.

Even so, the returns can be worthwhile. There is a quite substantial group of collectors who specialise in small or unusual books, some even in comics. At a country auction or school fête, it is not uncommon to pick up a bundle of old comics for 5 or 10 cents. Most of them will be worth only a few cents each, perhaps not even that. But if the bundle came out of a chest that had been stowed in an attic for twenty years, there could be treasure among the cheap rubbish. A 1963 *Avenger No 1* comic, for example, which cost one shilling the year it was published is currently valued at more than $500!

If this can happen with comics, then imagine the potential for old or rare books that have also been stowed away in an attic and only unearthed because Grandma had a "clean-out" for the local school fête. In amongst the penny dreadfuls that Mum used to read secretly in the garden shed (Gran thought they were unsuitable for young girls), could be a real treasure for the collectors' market. A first edition of Captain Cook's journal sold recently for $20,000!

Any book can be classified as a treasure, since, as with any treasure, its value or beauty is to a great extent in the eye of the treasure seeker. For example, an old diary written by a pioneer Great-Grandad would be considered a wonderful treasure by the family for purely nostalgic reasons. Historians would be delighted to obtain a copy of such a diary for the light it might shed on contemporary events or the everyday life of the period, and collectors would be anxious to get their hands on it for entirely different reasons again.

The secret of finding book treasures is twofold: knowing what to look for and knowing where to look. It goes without saying that becoming an expert on old books takes years of study and dedication. It is a very specialised skill, usually beyond the scope of amateur treasure hunters.

Neverthless, many books are often owned by people who do not appreciate their value and they are often buried among cheaper or less elegant works of "everyday" literature, which show up in crate-loads at auctions and fêtes. Or they are hidden on dusty shelves in second-hand bookshops and junk stores. And while it is true that there is a small army of professional dealers and collectors going through every bookshop and stall like literary locusts, there is no need for the amateur fossicker to despair. Just as a small nugget of gold can turn up in a seemingly barren stream that has been worked by fossickers for generations, so a book treasure can turn up in the most unlikely place.

So much for knowing what to look for, now the question is where to look. This is almost as important as knowing what to look for. Apart from the antique book auctions and shops, which are better left to the experts, the most rewarding places are old houses. This is not to say that Gran's bookcase and attic should be scoured at every family reunion, but it is worth keeping an eye open when visiting old homes.

Many collectors spend a great deal of time going through the book collections of old homes, for without television or radio, books were the main source of education and entertainment in past generations. Particularly in rural areas where, even until recent years, the electronic medium was poorly developed or provided only limited entertainment, books were the main, sometimes the only, source of relaxation. Over the years, many farmhouses and homesteads have acquired collections of books that would send a dealer into a frenzy of excitement.

In the city, particularly in modern families where relaxation and entertainment is catered for by TV, radio, movies and a daily avalanche of newspapers and magazines, books tend to play a less prominent role. While, fortunately, many books are still purchased and read, there is less tendency to keep them unless they have some specific quality or deal with a practical subject, such as cooking or child-rearing. There is usually less storage space in urban homes than in country houses, so city readers tend to have periodic "throw-outs" and there is less likelihood of old books gathering dust in the attic. This in turn means less rewarding fossicking grounds for treasure hunters.

However, what the cities lack in potential treasure grounds in the way of attics and barns, it compensates for with a never-ending run of fêtes, fairs, garage sales, and so on. So the scope for treasure seeking in the city, while perhaps not as exciting and romantic as in the country, is still potentially lucrative.

But the competition is fierce! Watch the bookstall in the first ten minutes of a well-publicised school fête. Like locusts, the experts, both

BOOKS TO BE TREASURED

Without the in-depth knowledge of the experts, amateur book hunters need to be aware of a few basic factors when searching for book treasures. Briefly, these can be listed as follows:

Age

Anything old falls into two standard categories: trash or treasure, and books are no exception. Broadly speaking, paperbacks degenerate quickly with age and mostly become tattered trash. However, paperbacks in good condition which have become rare over the years can be very valuable. The early Penguins are a typical example of this. Most bound books of quality become treasures with time. Generally, the older the book the more valuable it will be, although this will depend to a great extent on the other factors, not least its content. At all events, an old book should always be treated as treasure until its value can be determined.

Condition

Although the condition of an old book may not be as important as its content, nevertheless, as with all treasures, the better the condition, the higher the value. While discolouring of the pages due to age may make little difference to the value of an old book, missing pages reduce its value considerably. The condition of the cover is just as important. If the original cover is intact and in good condition, the better the chance of the book fetching a good price from a dealer. A leather binding will add value to any book. Dustcovers are important to the value of modern first editions. Books such as Ian Fleming's early James Bond series fetch high prices if they are in good condition and have retained their dustcover.

Edition

This is a very important factor, for first editions are prized by collectors, particularly of rare or valuable books. This does not mean that other editions are not valuable, simply that the first edition usually attracts more attention from collectors and is therefore more likely to bring a high price, though a second edition may be valuable for some specific reason. First editions of children's classics of 30 to 40 years ago are very popular with collectors. The Biggles books, and such classics as *The Magic Pudding*, *Winnie the Pooh* and almost anything written by May Gibbs can have considerable value — and these are just the sort of treasures that turn up at school and church fêtes.

Content

This is a grey area in valuing old or rare books since much depends on the market. In 1988 the value of old books related to Australia's past was greatly inflated by the nostalgia aroused by the bicentenary. A book on Tibet's past, by contrast, would have had impact only in a very specialised corner of the antique book market. Broadly speaking, fiction is less popular than factual content unless the novel is written by an author of note. A novel written in the 1930s by an unknown author would not have the interest of a book on railway trains of the same era. But a first edition of a Patrick White novel (around 1939) would be far more valuable than a book on fishing written in the same year. A book can be prized for its illustrations as much as its literary content. Specially illustrated editions of the Stevenson and Carroll classics fetch high prices in the collectors' market.

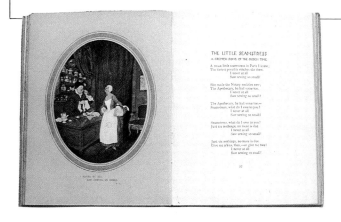

Old books are much prized as "attic" treasures, but they can be found almost anywhere. This *Picture Book for the French Red Cross* might be valued by a collector.

amateur and professional, scavenge through the piles, sifting and sorting with all the expertise of a mail-centre technician. Dealers, collectors, bargain book buyers, they all know what they want and move in fast. Smart treasure seekers are not far behind, for apart from getting in before the best is gone, watching and working alongside the experts can be an interesting and educational exercise.

PICTURES

Paintings, lithographs, drawings and other works of art can provide a fascinating and exciting form of treasure hunting. The same basic principles apply: leave the top-of-the-market stuff to the experts, most of whom dedicate a lifetime to studying and dealing in art. Concentrate on learning a few of the ground rules so that when a treasure is found, albeit only a modest one, it is not passed over. This may involve a course in art appreciation at a TAFE college or with one of the private enterprise colleges, or a close study of the many books available on the subject. Much depends on the extent to which you intend to hunt for art treasures. Attending exhibitions is a good way to learn the basics, and often such showings have excellent catalogues with considerable detail about the works on display.

Although there is always the possibility of coming upon an unidentified masterpiece, the chances are remote. But stowed away in attics, storerooms and garages all over the country there is a huge collection of art, much of which is not worth the canvas on which it is painted, but some of which makes worthwhile treasure-trove. As in all cases, the value is in the eye of the buyer, and while there are collectors and dealers who will scoff at anything that does not reflect an established school, there are those who will pay good prices for any form of art, be it traditional, art-deco, or modern.

The most valuable works are, of course, those by well-known artists, and the value will vary according to exactly which name is on the canvas as well as other aspects which create a demand for the picture. However, there are many dealers and collectors who will buy a painting by an unknown artist simply because they like it or because they feel that the artist has potential which might later be recognised, making the painting a worthwhile investment.

Since art in Australia has a relatively short history, the number of local artists who have achieved a name for their work is not great. Searching through the past treasures of a country homestead, it is always possible that an early "lost" work of one of these artists could be discovered. It needs no imagination to envisage the thrill of dusting off an early, unknown Hans Heysen painting, or a forgotten William Dobell. Although such finds are as unlikely as winning first prize in the lottery, it is the potential for such a lucky strike that provides the incentive for any form of treasure hunting.

More realistically, if far less exciting, is the likelihood of discovering pictures which, because of their age or their subject matter, will sell not so much as an investment, but as a collectors' item. People collect all manner of strange things, and enthusiastic collectors are prepared to pay large sums for an item that will complete or augment their collection. Oils and watercolours of "chocolate box" scenes, which grandmother put away years ago, can bring $100 and more from collectors of such items. Even picture postcards can fetch a good price. In 1986, the Museum of Australia paid $75,000 for a collection of 8000 early Australian postcards. The illustrations on the cards range from rare photographs of a Broken Hill strike to distinctive advertisements and cover the period from the turn of the century until the 1950s. How many such postcards turn up in school fêtes, in attic boxes, or in forgotten sea trunks!

by Guilielmo Blaeuw Dating from 1657.

Old maps are enjoying considerable popularity with some collectors. The Bicentennial year of 1988 revitalised interest in Australian maps.
Right: Knowledge of the subject is important in determining the value of old paintings.

COINS

Coins estimated to be worth more than $2 million are lost each year in Australia. Small wonder, then, that coins are among the most common treasures unearthed by treasure hunters. Because they are so widespread, they are often discovered when looking for a totally different treasure. Queensland artist Keith Courtenay, while on an expedition in the Palmer River goldfield, stumbled on a large Chinese pot containing 32,000 brass, copper, bronze and iron Chinese coins. All were old coins, since the pot had been buried during the gold rush, over 100 years ago. Some dated as far back as AD 618, and the early part of the Tang Dynasty.

Coins are found literally everywhere: in abandoned houses, ghost towns, wrecked ships, gutted cars, old workings, schools, parks, beaches, stockyards and down the backs of chairs. One enthusiastic treasure seeker, realising his children were reaching treasure-hunting age, organised an apprenticeship for them which initially involved looking solely for coins. Having found coins to a certain value in the house, they graduated to the garden, then to the neighbourhood and, finally, fully fledged in the art of treasure hunting, into the outside world. The fact that his entire brood satisfactorily completed their apprenticeships illustrates the amount of coin treasure that lies waiting to be found.

Apart from a keen eye, the most useful piece of equipment for finding coins is the metal detector. Coins are rarely buried very deeply, except perhaps on the beach, so the average metal detector, when properly tuned and set, reacts to even the smallest coins. In recent years treasure hunters have become a common sight along Australia's beaches in the late afternoon or evening, picking up the coins, watches, rings and personal belongings left by sun-worshipping crowds during the day. The bleachers at sports parks and stadiums after a big match are another good source of coin treasure. One shrewd operator, when working abandoned goldfields, leaves the mob to work the mullock heaps and tailings while he makes off in search of the site of an old toilet block, where, he claims, the miners lost a fortune in coins, personal belongings, and even small nuggets of gold. Dropping your pants, it seems, automatically empties the pockets!

While most of the coins recovered from parks, beaches or show-grounds are present-day currency and therefore only worth their face value, coins recovered from historic sites can be worth a great deal more. The Chinese coins found at the Palmer River goldfield would obviously be worth many thousands of dollars, while our ingenious toilet-block hunter might find coins that are at least 100 years old. A 1930 Australian penny in good condition was recently sold for $46,500, while a similar coin, as first issued by the Mint, fetched a world record price of $150,000.

While age is an important factor in determining the value of a coin, its rarity is critical. A coin which is fairly common, even though it is of good vintage, will not bring the price of a similarly dated coin of which there are only a few known to exist. Nevertheless, an old coin will always bring more than its face value, even a few years after it was issued, and particularly if it was a special issue. A 1985 proof set which was issued at a price of $40 would fetch around $75 on today's market. That is an average increase of about $25 a year, or more than 50 per cent — a good investment when compared with stock markets that crash as wildly as they did in 1987.

Australia's coin history is so complex it is a collector's dream. Bartering was the earliest form of trade, and coins were of little use, though many found their way ashore from ships of different nations berthed in Port Jackson. In 1791 the Spanish dollar, valued then at five shillings, was declared the legal currency for the colony, and in 1800 an English copper coin, the "Cartwheel" was introduced. It was valued at twopence. In 1812, in an interesting and unique development, Governor Macquarie had the centres of 4000 Spanish dollars cut out and reissued as separate coins — the ring, or "holey" dollar, and the centrepiece, which was called the "dump".

While few of these earliest Australian coins have survived — only 200 holey dollars are known to exist — there is always the possibility that more may be discovered as the result of a treasure hunt at some historic site. The value of such finds would be immense. Since those early days, different coinage has been used, mostly British, until Australia minted the first local coins in 1855. This was the decade of the gold rushes, so the sudden rise in the nation's population increased the demand for coins. The goldfields around the continent are probably the best place for treasure hunters to look for coins brought in by migrants seeking their own treasures, more than 130 years ago.

A typical arrangement of country farmhouse treasures on display in a city showroom.
PICTURE COURTESY MOSMAN COUNTRY ANTIQUES.

Above right: A fine example of country kitchen furniture, much in demand in trendy city suburbs.
PICTURE COURTESY MOSMAN COUNTRY ANTIQUES.

FARMHOUSE FURNITURE

It may seem odd to describe furniture as a treasure, but hunting for furniture can be not only a lucrative but an exciting pastime. The furniture is often old, but not antique in the accepted sense of the word, for dealing in antique furniture is a very demanding business requiring specialised knowledge and substantial amounts of money. The furniture which is the focus of a treasure hunt requires less knowledge and relatively small sums of money.

Even so, it is important to be able to recognise the finer points of the different styles, for with the high prices being paid for farmhouse furniture, it is not uncommon to be offered fakes and look-alikes, as well as shoddy restorations, instead of the genuine article. Treasure hunters should have sufficient knowledge to be able to tell the difference.

The returns may not be as great as one would expect from handling genuine antique furniture, but they can nevertheless be very rewarding. In any case, there are intangible rewards which to many treasure hunters are more important than the cash return. One of the more attractive features is that the transactions take place mostly in the country, in farms and homesteads where, apart from the relaxed environment, the people are warm and friendly. The stuffy, cloistered atmosphere of city or suburban antique dealers' rooms is far less enticing.

In the early days of settlement, most of Australia's better furniture was imported from overseas, and this is the area of most interest to antique dealers. Some of this furniture found its way into the homesteads of big country properties, but because of the cost and the problems of distance, isolated settlers often had to make do with a locally made product. Such furniture was often rough, with little character, and made from whatever timber happened to be convenient.

As time went on, however, a domestic furniture trade began to develop from these rude beginnings. The style of furniture was still not in the class of imported European products, but was more utilitarian, homely and suited to country living. When the Free Selection acts of the 1860s opened up the huge tracts of inland Australia to less wealthy landowners, the demand for farmhouse furniture increased rapidly and the trade came into its own. Made from good Australian timbers such as the prized red cedar (*Toona australis*) and many species of native pine, the furniture was designed to be practical and durable.

Today the pieces that were made to withstand the heavy duty of the farmhouse kitchen, living room and bedroom, are highly prized. They are enjoying a renewed popularity in the cities, where young families, especially, are enjoying the atmosphere of country furniture. Huge pine slab tables, dressers, blanket chests and old timber meat safes are setting off the kitchens of restored terrace houses and stone cottages in trendy suburbia. Milking stools and rocking chairs are back in vogue, to the extent that replicas are being mass-produced — which, of course, makes the genuine article even more valuable.

Treasure hunting for farmhouse furniture requires only a keen eye, good taste and an ability to bargain. It also requires a means of transport for the furniture, but this need be nothing more than a trailer, providing the furniture is well protected both from bumps and the weather.

Obviously, the farther removed from civilisation, the more potential a locality has for producing bargains. Country towns often have a second-hand furniture store and this is as good a place as any to start. What is a throw-out to a farmer's wife may be a treasure for a city kitchen, and the greater the bargain when buying, the greater the reward when selling.

Country auctions, which are held frequently in many areas, are the perfect place to look for furniture treasures. Some auctions are held at houses and homesteads where for some reason or other the furniture is being sold off. Others are held in local halls or showgrounds. The ordinary house or farm auction may involve sifting through a lot of rubbish to find a gem, but this very factor tends to discourage dealers to whom time is money, and the chances of a find are considerably better than at antique or specialised furniture auctions.

Then there are the farms and homesteads themselves, where many a bargain is only too gladly handed over by an owner who had been intending to get rid of it anyway. While not many people will part with their possessions without good reason, everyone enjoys a change of scene occasionally, and the right price might just persuade a reluctant seller that perhaps it *is* time to get rid of that old dresser and buy a new one! As so often happens, what is trash to one is treasure to another, and an old solid slab timber table that has been in the family for generations may have become an eyesore to the family that have eaten off it and scrubbed it for a lifetime but may add the perfect touch to a trendy kitchen in a suburban cottage. Persuasive bargaining may induce the owner to sell it happily for a few hundred dollars. A snap deal means the family gets money towards a new table, the trendy kitchen gets its country table, and the treasure hunter picks up an acceptable reward.

Experienced farm "fossickers" do not restrict their search to the kitchen or living room, of course, nor just to furniture, for there are often as many treasures to be found outside the farmhouse as inside. Hooped timber barrels, ploughshares and other ironwork, horse brasses and stone jars and bottles often lie unnoticed among the weeds and rubbish of the farmyard, just waiting for the discerning eye to fall on them. Since it is mostly rubbish to the farmer, he will be only too happy to get rid of it, and it is in such an environment that treasures are found and bargains are made.

STONEWARE AND BOTTLES

Like coins, old stoneware and pottery can be found in unlikely places since they once formed an integral part of everyday living. Stone jars were widely used for all kinds of domestic and commercial products and ranged in size from huge vats to small stone ginger-beer bottles. As glass and plastic gradually replaced the heavy stone for storing and carrying such products, stoneware was relegated to the rubbish heap. Treasure hunters forage through piles of junk or rubbish in sites that range from rubbish tips to old mining camps in search of pieces that have become valuable with time.

Stone ginger-beer bottles and whisky jars are not an uncommon find in the backyards of old homesteads or the ruins of abandoned houses. Some are worth little, since they are of relatively recent vintage, but some of the salt-glazed stone ginger-beer bottles that were "thrown" by convict potters in the early 1800s can fetch thousands of dollars. Even the later, mass-produced crown seal ginger-beer bottles will, providing they are well marked, fetch around $500, while the most humble stone bottle is likely to be worth $50, depending on its markings.

A common stone demijohn with more aesthetic than monetary value. Demijohns made by early Sydney potters can fetch up to $2000.

Stone demijohns, often known as "whisky jars", range in value from around $100 upwards, depending on the age, manufacturer and marking of the jar. A rare stoneware demijohn by the emigrant potter Thomas Field, who established his pottery in George Street, Sydney, in the 1840s, has been valued at around $1000, while a three-quart Bristol-glazed demijohn from the Lithgow pottery is estimated to be worth around $1700.

There are numerous other forms of stoneware which can, by virtue of their age or manufacture, fetch good prices at collectors' auctions. Kitchen jars, butter churns, bread crocks and urns are but a few of the treasures that were made in stoneware by the early potters, and which frequently lie unnoticed in the back of kitchen cupboards or among the weeds of back gardens. Even in the city, many old homes have valuable artefacts stowed away in dark recesses of dressers and cupboards. The squirrel mentality is not confined solely to those who live in the country!

Glass bottles are also found wherever there is a rubbish tip or a cupboard full of stored junk. Before the turn of the century, an English-styled bottle with a torpedo-shaped bottom was widely used in Australia. Designed by William Francis Hamilton, the bottle was designed so that it could never stand up. Hamilton determined that if the bottle were always lying down, the cork would always be moist and swollen, so that none of the contents would be lost though leakage or evaporation.

The Hamilton bottle was in vogue throughout the latter half of the nineteenth century, so there are many still in existence. They are a fairly robust type of glassware and it is not uncommon for treasure hunters to find one. Depending on their vintage and marking, a torpedo bottle, as it was known, could fetch between $150 and $800 in a collectors' market, although the more common, unmarked bottles might be worth less than $10.

Codd bottles, named after their designer, Hiram Codd, took over from Hamilton's torpedo bottles at around the turn of the century, being particularly popular with soft-drink manufacturers. The Codd is a flat-bottomed bottle with conventional shoulders into which a restriction has been moulded to enclose a glass marble seal. This seal avoids the need to lie the bottle on its side to keep the cork wet. Codd bottles are still dug up in back gardens and are relatively low in value unless they have some distinctive marking.

Many of the bottles recovered from Australia's past are manufactured overseas, and it would take a catalogue the size of this book to detail them all. Some of the more interesting varieties carried spirits such as whisky, gin and brandy. Square "Schiedam" gin bottles with ornate mouldings are typical of the type of imported bottle that has some value in a collectors' market. These black or dark green square bottles can range in value from $50 to $1500. Broken pieces of old liquor bottles are often unearthed at abandoned goldfield sites.

BRIC-A-BRAC

Under this heading can be included all the other possible treasures that might be found during a treasure hunt. Far too numerous to detail, they may range from a pen nib to an Aboriginal axe-head and may be found on sites as diverse as a mountain top and an offshore reef. They will almost certainly have one thing in common — they are relics of the past, for age is mostly the key factor in creating a treasure. Even gold and gemstones, seeing the light of day for the first time as they are extracted from the soil, are the product of age, for their foundations lie in the creation of the earth, millions of years ago.

Amateur treasure hunters will find that almost anything that is not of recent manufacture has some value. The following list is not complete; it is hard to imagine that such a list could be complete. Rather, this is a list of treasures, some with monetary value, some with instrinsic value, some with aesthetic value, to complement those described in more detail earlier in this chapter. It is a list from which the budding treasure seeker can determine which treasure might form the focus of the family's next expedition:

Aboriginal artefacts
Advertising boards, labels, etc.
Ashtrays
Badges
Bedwarmers
Bells
Brassware
Bricks (convict)
Ceramics
Cigar boxes
Cigarette cards
Clocks
Coal scuttles
Copperware
Crockery
Cutlery
Dolls
Farm implements, memorabilia
Fire screens
Flat irons
Fossils
Glassware
Guns
Horse brasses
Ink bottles and wells
Jewellery
Keys
Kitchenware
Lamps
Locks
Maps
Matchboxes, holders
Medallions
Name plates
Pewter
Photographs
Pincushions
Pipes
Pokers
Postcards
Posters
Pot lids
Railway memorabilia
Shells
Ship's memorabilia
Signs
Silverware
Stamps
Taps
Tobacco boxes
Tins
Toys

FURTHER READING

GENERAL

The Australian Encyclopaedia
Grolier (1979)

Ghost Towns of Australia,
George Farwell (1965)

Lost Treasures in Australia,
Kenneth W. Byron (1964)

Tasmania,
R. J. Solomon (1972)

GOLD AND GEMSTONES

Anakie,
Walda L. Scholler (1985)

Australian and New Zealand Gemstones,
Bill Myatt (1972)

Australian Rocks, Minerals and Gemstones,
R. O. Chalmers (1967)

Collecting Australian Gemstones,
Bill James (1970)

Detecting Australian Treasure,
H. K. Garland and D. J. Jones (1982)

Fossicker's Guide to Gold, Minerals and Gemstones of South Australia,
Len Dallow (1983)

Fossicking Central Queensland Sapphires,
Department of Mines, Queensland

Geology and Mineral Resources of Tasmania,
I. B. Jennings and E. Williams

Gold,
T. E. Johnstone (1980)

Gold at Jupiter Creek,
Department of Mines and Energy, South Australia

Gold Fossicking,
Department of Mines, Queensland (1987)

Gold in South Australia,
Department of Mines and Energy, South Australia

Gold in Western Australia,
Western Australian Department of Mines (1985)

Gold Prospecting,
Douglas M. Stone (1977)

Guide to Australian Gemstones,
Reader's Digest

Lost Mines and Treasures of Northern Australia,
F. G. Brown (1983)

Minerals in Victoria,
Victorian Department of Mines (1970)

Occurrences of Gemstone Minerals in Tasmania,
Tasmanian Department of Mines

Opal in South Australia,
South Australian Department of Mines
and Energy (1986)

Panning and Prospecting for Beginners,
H. K. Garland (1975)

Prospector's Guide,
Department of Minerals and Energy, Victoria (1980)

SHIPWRECKS AND HIDDEN TREASURES

Ghost Ports of Australia,
Jeff Toghill (1986)

The Mahogany Ship,
J. K. Loney (1974)

Sea Adventures and Wrecks on the New South Wales Coast,
J. K. Loney

The Secret Discovery of Australia,
Kenneth Gordon McIntyre (1977)

We Discovered an Island,
Stephen Yarrow (1980)

COLLECTABLES

Australiana & Collectables Price Guide,
David Westcott (1986)

Collecting Australia's Past,
Douglas Baglin and Frances Wheelhouse (1981)

Collectors Annual Bicentennial Edition,
(1987)

Guide to Collecting Australiana,
Juliana and Toby Hooper (1978)

Treasure Hunters Manual No 7,
Karl von Mueller (1979)

Index